James E.H. Gordon

Four Lectures of Static Electric Induction

Delivered at the Royal Institution of Great Britain, 1879

James E.H. Gordon

Four Lectures of Static Electric Induction
Delivered at the Royal Institution of Great Britain, 1879

ISBN/EAN: 9783337068950

Printed in Europe, USA, Canada, Australia, Japan

Cover: Foto ©berggeist007 / pixelio.de

More available books at **www.hansebooks.com**

FOUR LECTURES

ON

STATIC ELECTRIC INDUCTION.

BY

J. E. H. GORDON, B.A.,
Assistant Secretary of the British Association.

DELIVERED AT THE

ROYAL INSTITUTION OF GREAT BRITAIN, 1879.

Illud in his rebus non est mirabile, quare,
omnia cum rerum primordia sint in motu,
summa tamen summa videatur stare quiete,
praeterquam siquid proprio dat corpore motus.
omnis enim longe nostris ab sensibus infra
primorum natura jacet.
LUCRETIUS, II. 308.

New York :

D. VAN NOSTRAND, PUBLISHER,
23, WARREN AND 27, MURRAY STREETS.

1881.

CONTENTS.

LECTURE I.

Thursday, *January 16th*, 1879.

LECTURE II.

THURSDAY, *January* 23rd.

LECTURE III.

THURSDAY, *January* 30th.

If induction is a state of strain of the medium through which it is propagated, different

LECTURE IV.

THURSDAY, *February 6th.*

FOUR LECTURES

ON

ELECTROSTATIC INDUCTION,

DELIVERED AT THE ROYAL INSTITUTION,

Jan. 16, 23, 30, and Feb. 6,

1879,

BY J. E. H. GORDON, B.A.

(Assistant Secretary of the British Association).

LECTURE I.

Jan. 16.

Introductory.

" Amongst the actions of different kinds into which electricity has conventionally been subdivided, there is, I think, none which excels, or even equals, in importance that called *Induction*. It is of the most general influence in electrical phenomena, appearing to be concerned in every one of them, and has in reality the character of a first, essential, and fundamental principle. Its comprehension is so important that I think we cannot proceed much further in the investigation of

B

the laws of electricity without a more thorough under-
standing of its nature. How otherwise can we hope
to comprehend the harmony and even unity of action
which doubtless governs electrical excitement by friction,
by chemical means, by heat, by magnetic influence, by
evaporation, and even by the living being?"

So, forty-two years ago, wrote the Master
whose memory is honoured wherever the
study of natural laws is loved, and whom
in this place we should more especially
remember, as the Royal Institution was
his home and workshop during all the best
years of his life. Need I add that the
passage I have just read is from the " Experimental Researches " of Faraday?

The subject of our study to-day and in
the other lectures of this course will be
such of the laws of induction as are now
clearly known. I shall first endeavour to
show you what the term induction means,
and what is the problem about it which for
fifty years students of nature have been
trying to solve. The problem is partly

solved now, but much remains to be done. It will, I think, be pleasant to follow the stages of discovery up to to-day, and perhaps to look a little forward and try to see what the discoveries which may some day be made may lead to.

First, we have the simple phenomena of induction and electrification, some of which I shall shortly show to you.

The unsolved question, "What is electricity?" we shall not attempt to touch upon. It is sufficient for our purpose to know that when a body exhibits certain properties it is said to be electrified, or to be in a state of electrification. We also know how to produce this state at will, but we know next to nothing of its nature. We do not know whether the properties of an electrified body are caused by one or two electric fluids entering it or leaving it, as water into a sponge; or by a motion of its molecules, as when a body is heated;

or by a strain or twist of its structure, as when steel is magnetized.

We have no conception of electricity apart from the electrified body; we have no experience of its independent existence. Let us now begin the study of certain phenomena of electrification which it is necessary for us to understand before commencing the study of induction.

If we rub a piece of sealing-wax or glass with a silk handkerchief we find that it has the power of attracting light bodies, as you see. The glass or sealing-wax, after being rubbed, is found to be in the state called " electrification."

I must ask your pardon for repeating so well known an experiment as this, but my reason for doing so is that I wish to call your attention to a feature in it which usually receives but small attention. The point I want you to notice is that when I hold the electrified body near these light paper shavings, the action takes place from the sealing-wax *across the intermediate air.* What is the nature of the action? We do not fully know yet, but it is called *Induction.*

This is only one form of induction, but I shall hope to show you others. Meanwhile, note the following definition to begin with.

Every electrified body from which no electrification is allowed to escape has a particular action on all neighbouring bodies, and this action is called induction.

Before we go any further in the study of induction we must inquire, Is there any difference between the electrification produced by rubbing sealing-wax and that produced by rubbing glass? We may answer at once that there is, and the difference is a very curious one, namely, that the properties of the two kinds of electrification are exactly opposite to one another.

By opposite, I mean this : if by any means
equal quantities of the two electrifications
be added together, they will exactly neutralize each other, or, in other words,
adding a quantity of one kind of electricity
is the same as taking away an equal quantity of the other. There are a great many
ways of producing electrification, but all
electrification is of one or the other kind—
either that of glass or that of sealing-wax.
For convenience, glass electricity is called
positive, sealing-wax electricity, negative.
Here I have an electric machine, which is
simply a convenient arrangement for rubbing glass and silk together.

I have shown you one form of induction,
namely, the attraction of light bodies by
an electrified body. Let us now examine
the effect of electrified bodies on each other.
Here (Fig. 1) I have some pieces of sealing-
wax and glass, and a means of suspending
any one of them. We find that sealing-

wax repels sealing-wax, glass repels glass, glass and sealing-wax attract each other, or, generally, like electricities repel, unlike attract. Hence you see that there is always a force acting between electrified

FIG I

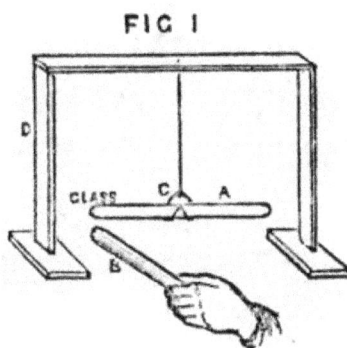

bodies, and that, when the electrifications are alike, it is a repulsive force, when different, it is attractive.

This force acts through the air or *other substance* between the electrified bodies, as you see when I put this plate of paraffin-wax between the suspended rod and the one in my hand. It is, therefore, a form of induction. What is the machinery that

conveys this force across the air or paraffin?
This is a question to which we shall attempt
to give a partial answer later on.

Conductors and Insulators.—In certain
substances, such as metals, electrification
is able to move freely; that is, if one end
of a metal rod receives electrification, the
electrification is at once *conducted* to every
part of it, as you may here see. (Fig 2.)

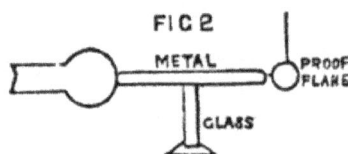

FIG 2

The proof plane being applied to the
end furthest from the machine, it is found
to be electrified. These substances are
called conductors. In other substances
electricity is not able to move freely, and
if one end is electrified, the other remains
in an unelectrified state, as may be seen by
substituting a glass rod for the metal one
in the preceding experiment. The electricity

being *insulated* at the first end, these substances are called Insulators. It is possible
to insulate a conductor by placing it on a
glass stand, and we can then study the
movements of the electricity in it without
the latter being able to escape.

We are now in a position to study the
effect of the induction of an electrified body
upon a conductor near it. Here is our insu-

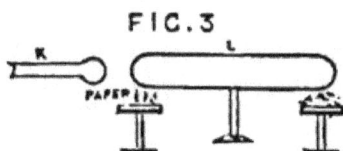

FIG. 3

lated conductor L, and we will place it near
the charged knob K of the electric machine.
(Fig. 3.) The jumping up of these paper slips
shows that the metal is electrified, but no
electricity has passed to it, and it is electrified by induction. Thus we learn that by induction an electrified body electrifies bodies
in its neighbourhood. This metal rod is
electrified, although no electricity has been

transferred to it. Let us now examine what difference there is between the electrification produced in this rod by *conduction*, or direct transfer of electricity, and by *induction*. First let us electrify the rod by conduction, and let us, after stopping the machine, test both ends with the proof plane. We see that both ends are *positively* electrified. We have stopped and discharged the machine, but still the electricity remains in the cylinder, and will remain there until some *conducting* path is opened for it, as by touching it with the finger.

FIC 4

The actions are just as if a portion of an electric fluid had been forced from the machine to the cylinder, had distributed itself all over the latter, as water finds its own level, and had now been drawn off through my body as water through a pipe.

Let us now return and examine the parallel induction phenomena. We see by the attraction of the papers that the cylinder is electrified at both ends. Let us now examine what kind of electricity there is at each end. We find, first, that the end nearest to the machine is negative, the far end positive. We now see why electrified bodies attract light objects; they first induce on the side nearest to them an electrification opposite to their own and then attract it.

Let us now stop the machine. All signs of electrification disappear. Thus we see, when electrification is produced by induction, there is nothing analogous to the transfer of fluid from the machine to the cylinder. It is more as if, to use a bold simile, by some straining force, the cylinder was distorted into an electrified state, and,

just as when we distort a plastic substance,
any increase of length is accompanied by a

FIG 5

corresponding decrease of thickness, so that
the volume remains the same, so, when we
electrically distort this cylinder by induction, every increase of electrification at one
part of it, that is, any appearance of positive electricity, is accompanied by a decrease
of electrification at another part, that is, an
appearance of negative electricity.

The most rigorous and accurate experiments have shown that these two quantities, viz., the increase and decrease of
electrification in a body, when acted on inductively, are exactly equal. I can show you
a rough experiment to illustrate this point.

I have here two gold leaf electroscopes,

which are exceedingly delicate machines, for detecting small quantities of electricity. We now again electrify the cylinder by induction, and by means of the proof plane transfer

FIC.6

a little electricity from each end to the electroscopes respectively. We now stop the machine and remove the cylinder. Each electroscope remains charged, one positively and one negatively, and each with a

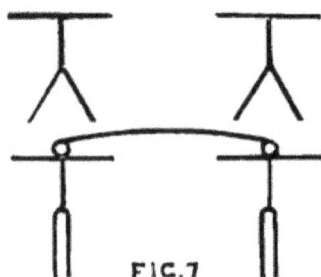

FIC.7

charge whose strength is proportional to the induced charges on the two ends of the cylinders respectively. If these charges

are equal, they should neutralize each other
when I connect the electroscopes (Fig. 7),
and you see they do so.*

The reason why it is so important that
we should see clearly that equal quantities
of both kinds of electrification are always
produced by induction is that this experimental fact shows us that the action of
induction is to produce something analogous
to a distortion of the electrified body, and
that, if this were not the case, but a greater
quantity of one kind of electricity than
another was induced, it would show that
something had been added to or taken from
the induced body, and the action would be
more analogous to a change of bulk than to
a distortion of molecular shape.

The problem, then, that we have before
us is: "Given the known experimental
facts which we have just been considering;

* The difference of distribution at the two ends is
not sufficient to affect this experiment.

given that there is an action, which we call induction, across air and other insulators from an electrified body to other bodies in the neighbourhood ; that the induction causes these attractions, and repulsions, and 'inducings' of electrifications which we have spoken of, what is the machinery by means of which this induction acts? What is the nature of the lever, the rope, or the pushing pole, which strains, and pulls, and pushes across the air, or glass, or other non-conductor which we place between the induced and inducing bodies?" We must attempt to answer this question bit by bit, and our first attempt shall be based on the difference between Induction and Conduction.

We have seen that when a piece of glass or other insulator is placed in contact with the conductor of an electric machine, it is thrown into a state of strain and distortion, but that the electricity does not

escape through it. When, however, a
metal or other conductor takes the place
of the glass, there is no appearance of such
a state of strain at all. What is the explanation of this? It is this. Equally in
conductors and insulators a state of strain
occurs, but in conductors *this state of strain
is continually giving way,* while in insulators
it does not do so. To keep up the state
of strain in a conductor would be as difficult as to keep up a pressure of steam in a
boiler with a large hole in it.

Let me show you a mechanical experiment in illustration—only in illustration,
remember, not in explanation—of what I
mean. Here is a vessel, U, connected to
the water-pipes at one end and to a pressure gauge, S, at the other. There is no
escape for the water, it cannot flow or
move, and the gauge shows a considerable
pressure. I now turn the tap T, and allow
a stream of water to escape. The pressure

and strain is relieved and the gauge falls;
that is, as soon as the state of constraint
gives way and the current flows, it is seen
that the strain no longer exists.   In the

FIC.8.

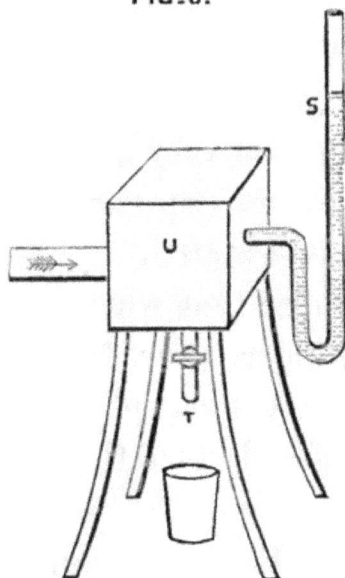

analogous electrical case, bodies in which
the state of constraint easily gives way do
not show the phenomena of strain or induction, but allow the electricity to flow
freely, and these are called conductors;
while, on the other hand, bodies which

have a great power of resistance to the
straining force can be greatly strained
without allowing a current of electricity to
flow. These are called insulators or non-
conductors. When such a body is subjected to a powerful straining or inducing
electric force, it exhibits the phenomena of
strain or induction very strongly. Let me
now show you an experiment illustrating
what I have just stated.

You remember that when we placed the
insulated cylinder near the machine the
induction which took place charged the
near end negatively and the far end positively. In this experiment we are only
concerned with the near end, and we will
lengthen our cylinder so as to get the far
end out of our way. How are we to do
this ? This is a large room, and no doubt
we might, at some considerable trouble and
expense, so lengthen the cylinder that we
could remove its other end to a distance of

some 20 or 30 feet.   But we can do better
than that.   We will make the whole world
part of our conductor.   The earth, owing
to the water in it, is a good conductor.
We will connect this wire from the cylinder
to the water pipes, and now (Fig. 9) we
have one end of our conductor on the table
and the other safely out of our way some-
where in Australia.

Now you see there is air in the space

FIG.9.

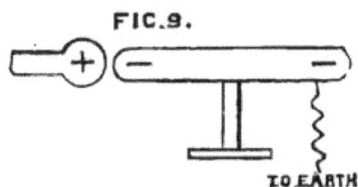

TO EARTH

between the machine and the cylinder,
and the state of strain will commence as
soon as I begin to work the machine, as
you see by the divergence of the gold leaves
of the electroscope when I take the proof
plane from the cylinder to it; in other
words, electricity is induced on our end of
the conductor.

c 2

This induced electricity will remain here during the action of the machine as long as the air or other insulator is between the conductor and the cylinder ; that is, as long as the substance between the conductor and the cylinder resists the straining force, so long will the state of strain be kept up. If, however, I connect them by some substance which offers an exceedingly small resistance to the straining force, as this metal bar, the state of strain at once gives way, and all induction ceases, and no divergence of the electroscope can be obtained. The electricity at the same time flows away, and distributes itself in the earth. This experiment is somewhat ana- logous to the mechanical one I showed you, where the strain was relieved by opening a tap and allowing the water to flow away.

The particles of glass move more freely over each other when hot than when cold, and hence we should expect that hot glass would yield more easily to a straining force than cold glass would. The following experiment shows that this is the case. Here is a glass flask containing mercury,

FIG. 10.

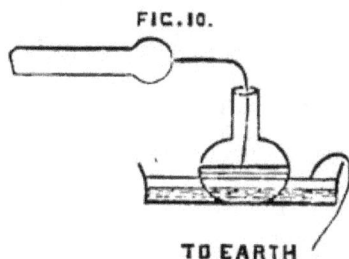

TO EARTH

and set in a dish of mercury. The mercury inside is connected to the electric machine, and that outside to the earth. On working the machine it is found, first, that no electricity can escape through the flask; secondly, that there is a strong induced charge on the mercury outside. Now let the mercury be made hot. It heats the glass, the particles move more freely over each other, the glass yields to the straining force, electricity escapes through it, and at the same time all induction ceases.

We have, in this lecture, by various
means produced electricity, and we have
produced sometimes one kind, and sometimes the other. It is important to examine whether we can actually produce one
kind alone. If this were possible, we might
actually increase the quantity of electricity
in the world. Experiment shows us that
we cannot do this. For every bit of
positive electricity that we produce we produce an exactly equal quantity of negative.
We cannot make or destroy electricity ; we
can only strain bodies so that their two
ends shall show opposite electrical properties. When we rubbed glass we produced positive electricity on its surface.
Was not that a creation of electricity? No ;
for an exactly equal quantity of negative
electricity was produced on the rubber, as
I can show you. (The rubber, on being
laid on the electroscope, caused a strong
divergence of the leaves.)   To show that

this negative is equal to the positive, a very simple experiment will suffice. I rub this sealing-wax till, by the cracking, you can hear that it is highly electrified, but do not remove the rubber from it. You see there is no effect on the electroscope. The reason is that the action of the positive on the rubber exactly balances that of the negative on the sealing-wax.

In the electric machine itself equal quantities are produced, only the rubbers are connected to the earth, so the negative escapes, and only the positive is kept. Here is the machine placed on an insulating stand, and a wire from the rubber brings the negative as well as the positive to the conductor. I work the machine, and you see even the gold leaf electroscope shows no sign of electrification. This shows that equal quantities are always produced. But when we rub sealing-wax and silk, and remove the silk, we have to all appearance negative alone in the wax. No ; for the instant the balancing positive is removed, the negative, by induction, produces a fresh positive on all surrounding bodies. It does so then, and not till then, as we may see by making one of those neighbouring bodies the electroscope, while the rubber, which has been removed, induces negative on bodies near it. No electrification of one kind only can be produced anywhere. If we charge a balloon, and send it up as high as possible, it will still induce an opposite charge, whose total amount will be equal to its own charge, on whatever is nearest to it, be it earth, clouds, or clear air. We have no means of knowing how much or in what way the earth itself, with its atmosphere, is charged ; but this we know, that, whatever its charge may be, it will induce an exactly equal opposite one on the moon, the sun, and even the most distant stars.

LECTURE II.

JAN. 23.

THE LEYDEN JAR, INDUCTION IN CURVED LINES.

TO-DAY we will continue our inquiry as to
the reasons for supposing induction to be
a state of strain, and we will now attempt

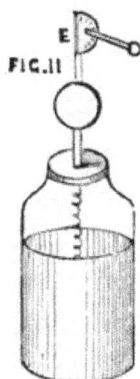

FIG. 11

to obtain an answer to this inquiry from a
study of the various phenomena exhibited
by the instrument known as the Leyden
jar.

The Leyden jar, in its most common form, consists of a wide-mouthed bottle, coated inside and out with tinfoil. The wooden stopper supports a brass knob, which communicates, by means of a wire or chain, with the inside coating. In order that the inside and outside coatings may be well insulated from each other, they do not reach quite to the top of the jar. Thus the jar forms a system of two conductors (the tinfoils), separated by a thin insulator (the glass). If we connect the knob of the jar to the machine and work the latter, we can charge the inside tinfoil, and, on removing the machine, this tinfoil will retain its charge for a considerable time, as is shown by this electroscope E on the knob. (Fig.11.)

This insulated electrified conductor now acts by induction through the glass of the jar, and induces electricity on the outer tinfoil conductor. As long as the jar is insulated there will be negative electricity

on the nearest portion of this outside conductor—that is, the inner surface of the
outside tinfoil—and positive electricity on
the further or outer surface.   (Fig. 12.)

Now, let us, as before, remove the further
end of the outer conductor to the other

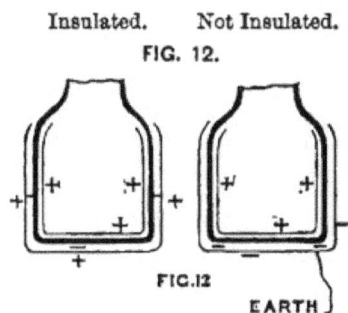

Insulated.   Not Insulated.
FIG. 12.

side of the world by connecting the outer
coating to the water-pipes ; we shall then
have the whole of the outer tinfoil negatively electrified.   Here, then, we have our
two conductors oppositely charged, acting
on each other inductively through the
glass.

Some idea of the intensity of the strain
may be obtained by " discharging " the

jar, as it is called. I have here a pair of
what is called " discharging tongs," which
consist of a conveniently-shaped conductor
fixed to an insulating handle. I hold the
"tongs" so that one knob touches the outer
conductor of the Leyden jar (Fig. 13), and

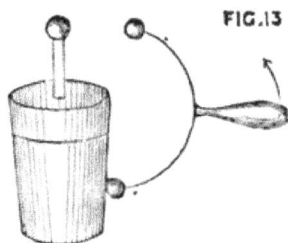

FIG. 13

then bring the other knob of the " tongs "
near the knob of the jar, which, we remember, is connected to the inner coating.

The strain between the conductors is now
taking place through two different insulators; that is, first through the glass of the
jar, second, through the air between the
two knobs, viz., the knob of the jar and the
upper knob of the tongs. The glass is

strong enough to resist the straining force of such a charge of electricity as the jar now has. We now bring the knobs nearer together. The straining force across the air between them gets greater and greater, and, at the same time, as the thickness of the air diminishes, its power of resistance, or of sustaining the state of strain, gets less and less, till here, as you see, the air breaks and gives way, and the electricities rush together with a flash and a report.

The straining force of the charge which we gave to the inner coating is removed with the charge, and immediately after the flash and discharge the falling of the electroscope shows that there is no electricity whatever on either coating of the jar. But, see, now the electricity seems to be returning; the slight motion of the electroscope ball shows that a slight charge has returned to the inner coating. On applying the

discharging tongs a spark occurs as before,
only it is very much feebler, and the jar is
now completely discharged.

What does this mean? Where did the
second charge of electricity come from?
Let me show you a mechanical experiment,

FIG.14

which will help us to an explanation. I
have here a strip of gutta-percha, of which
the lower end is fixed to a block. As it is
somewhat small, I will turn it edgeways to
the lime-light, and project the shadow of it
on the screen. (Fig. 14.) I now bend it
down by my finger, and suddenly let it go.

It flies up nearly, but not quite to the vertical position, rests an instant, and then moves slowly on till it is quite vertical.

If a spring index had been applied to it, it would have been seen that while pressed down it exercised a strong upward pressure. At the moment when it was at rest a little way from the vertical it would be exercising no pressure, and then it would be seen that as it began to again move towards the vertical, it would again exercise pressure. The gutta-percha was strained or distorted by the finger. When the straining force was removed, the strain, suddenly, nearly disappeared, but not quite. Then, in the course of the next few minutes, the disappearance of the strain or distortion was completed slowly.

The electrical case is exactly analogous. The pressure of the finger represents the first charging of the inner tinfoil; the straining of the gutta-percha represents the electrical straining of the glass. The pressure on the finger by the strained india-rubber represents the induction on the outer conductor. As in the gutta-percha, when the straining force is removed, the strain or distortion nearly disappears, and the upward pressure exercised by it entirely ceases, so in the Leyden jar, when the inducing electricity is taken away, the strain of the glass almost vanishes, and the induced charge disappears.

The strain or distortion of the glass, however, has only almost, but not entirely, disappeared; and now that there is no straining force interfering, the particles of the glass move over each other slowly, and in the course of a few minutes return to their normal state. But now, while the inner conductor has remained insulated, a change has occurred in the electrical arrangement of the particles of glass adjoining it. The state of strain has altered. They have changed from a more to a less distorted shape.

Now, in the ordinary phenomena of induction, what happens when we alter the state of strain of an insulator by bringing a charged body near it? Why, it induces electricity on any adjoining conductor. Similarly in the present case, when the elasticity of the glass brings it from a more to a less strained state, and so alters the state of strain, a charge is produced on the insulated conductor, and this is the residual charge which we have been inquiring about.

We notice that this residual charge returns slowly and gradually. Now, when a body is mechanically distorted, and is returning to its normal state by virtue of its elasticity, anything which enables the particles to move more freely over each other, such as tapping or jarring, will hasten that return. If, for instance, we have a heaped

D

up tray of sand slowly returning to its normal unstrained state of being level under the action of gravitation, any tapping of the tray will hasten the recovery from the state of strain; that is, hasten the return of the surface to a level state by enabling the particles to slide more freely over each other.

Now, if our supposition that these induction phenomena are the effects of strain, and that the residual charge is the returning of the distorted particles of glass to their normal state, is correct, any tapping or jarring of the glass should hasten this return; that is, hasten the appearance of the residual charge. In the *Phil. Trans.* for 1876, Dr. Hopkinson has shown that this actually occurs, and I shall now hope to repeat his experiment before you.

For this purpose we will not be content with our electro*scope*, but, as we wish to measure electrification, we must use an

electro*meter*. The instrument here is called
the quadrant-electrometer, and is the invention of Sir William Thomson.

Here is the instrument in the simple form
made by Messrs. Elliott, and here (right-hand
half of Fig. 15) is a diagram of the essential
parts of it. A sort of brass pill-box, supported horizontally, is cut into four quarters

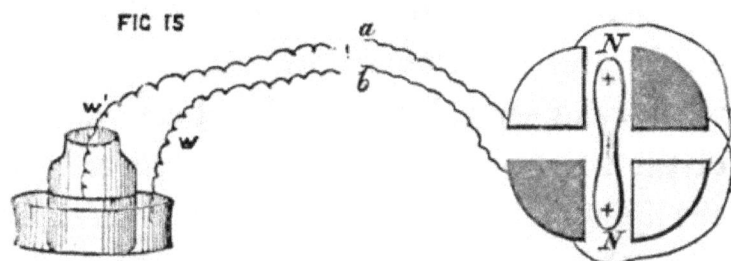

FIC IS

or quadrants, each of which is insulated from
the one next it, but connected to the one
diagonally opposite to it. An aluminium
needle, N N, is suspended, so that it can
swing like a compass needle inside the
pill-box. The needle has a strong positive
charge. When wires from the inner and
outer coatings of a Leyden jar are con-

nected to the wires a and b respectively, so that the unshaded quadrants are positive and the shaded ones negative, it will be seen that the action of all four quadrants is to turn the needle in the direction of the hands of a watch.*

The motion of the index needle itself is very small, but attached to it is a small mirror. Light from this lime-light falls on it, and is reflected on to the screen where you see this spot. The least motion of the needle and mirror of course moves the light-spot along the screen. The amount of motion is noted by means of the scale attached to the screen.

* The instrument can also be used to adjust two similar electrifications to equality, for if a and b are both positively electrified, the quadrants will tend to turn the needle in opposite directions, and it will go to the right or left according to which electrification is strongest. When we have varied one of the electrifications till there is no deflection, we know that we have made them equal.

Our Leyden jar in this case is made in a form somewhat different to that which we have been considering. The insulator is, as before, a small glass bottle,* but the conductors consist of strong sulphuric acid. Some is put inside the jar, and some in a glass dish in which the jar is set. To charge the jar a platinum wire comes from the electric machine to the acid inside the bottle, while that outside is connected to earth.

The jar is charged for two or three minutes and then discharged. I now connect the inner and outer coatings for about a minute by holding the wires from them together between the finger and thumb. The wires are now separated and connected to the quadrants of the electrometer, the earth connection being removed. The spot of light showing the motion of the

* A bottle about four inches high, intended for making a small Leyden jar, was used.

needle at once begins to move along the scale at the rate of about three inches per second, showing that the residual charge is coming slowly out.

On tapping the edge of the bottle briskly with a piece of hard wood, the pace at which the light-spot moves is at once trebled,* showing that while tapping is going on, the residual charge returns much more quickly.

If we wish to repeat the experiment, we can discharge the electrometer, and bring the light-spot back to zero by holding the wires together for a moment between the finger and thumb.

Thus we see that any mechanical vibration communicated to the particles of glass increases their freedom of motion among each other, and, therefore, enables them to recover more quickly after they have been

* This experiment was plainly seen by a large audience.

thrown into a state of strain by electric induction.

This experiment is nearly conclusive, I think, as to induction being a state of strain. If the charge of a jar were caused by an action at a distance, we should have to state that part of an action at a distance gets entangled in the glass and left behind, and that tapping helps it to escape. This hardly sounds probable.

This concludes what we have to say about the Leyden jar. We have shown that induction is a state of strain. We will now begin to inquire into the nature of this strain, and try to find out a little about how it is propagated from place to place.

Is it propagated simply in straight lines? does the inducing electricity stretch out an arm through the insulator, and pull at the second conductor? or does it act only on the particles of the insulator which are nearest to it, leaving it to them to act on the next set, and so to carry on the strain from particle to particle till it arrives at the second conductor?

Faraday asked himself the question, and it occurred to him that there was a very simple method of arriving at an answer. If the induction is propagated from particle

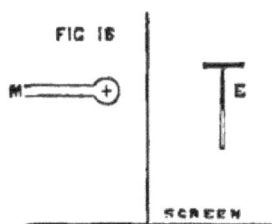

to particle of the insulator, it can travel along any direction where there is a continuous chain of insulating particles, whether this chain forms a straight line or not; in other words, it can *turn a corner*. If it were a " direct action at a distance " (whatever that may mean) it could only travel in straight lines.

The following experiment is a modifica-
tion of one designed by Faraday to show
that induction can take place in curved
lines. No induction can take place through
a metal screen which is connected to earth.
The simplest way to prove this will be to
try an experiment. Here (Fig. 16) is a
large metal screen connected to earth, and

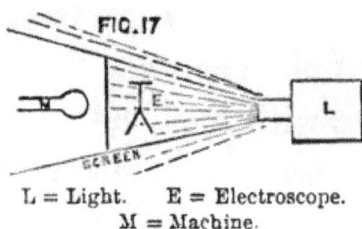

FIG.17

L = Light.   E = Electroscope.
M = Machine.

I place the electric machine on one side
of it, and the gold leaf electroscope on
the other. However strongly I work the
machine, there is no divergence of the
leaves. I now (Fig. 17) take away this
screen and put in a smaller one. In
order that we may be sure that this,
though smaller, is still large enough to cut
off all straight lines from every part of the

machine to the electroscope, I put the lime-
light as you see (Fig. 17.)   You see that
the optical shadow of the electroscope falls
entirely on the screen, and the shadow of
the screen entirely covers the machine·
On working the machine the leaves diverge
widely.

How did the induction get to them?
Our experiment with the large screen shows
that it could not have passed through the
screen; that with the lime-light shows that
it could not have come in a *straight* line
past the edge of the small screen; and,
therefore, we see that it must have come
in *curved* lines round the edge of the small
screen.

This experiment shows a point of difference between the projection through air of
light and of electric induction, for while
the edge of the optical shadow is almost
on the straight line from the source of
light through the edge of the screen, the

electrical shadow does not extend nearly
so far, but the induction curves considerably
round the edge of the interposing screen,
and extends in every direction, in which
there is a continuous chain of insulating
matter.

I think the experiment which we have
just tried will prove to you the existence
of induction in curved lines. I will now,

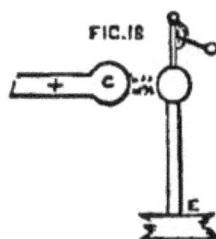

however, endeavour actually to show you
a curved line of induction. If I connect
an electroscope to a knob placed near an
electric machine and connected to earth,
you remember that, when we work the
machine so that sparks pass slowly, the
electroscope shows a strong induced charge,

which increases to a maximum just before
each spark and immediately after the spark
falls to zero. (Fig. 18.) I will repeat the
experiment, as we have not seen it exactly
in this form.

This experiment shows that induction
precedes discharge. All that we know
about the subject shows that this is uni-
versal law, namely, that there must always
be induction along the whole path between
the conductors before discharge can take
place. It is clear that this law ought to hold,
for discharge is only the sudden breaking
down of a state of strain, and there can be
no breaking down of strain except where
strain exists, and induction is strain.

The fact of a spark passing along any
path shows that induction was previously
taking place along that path. It does not,
however, show that the whole induction
was along that particular path even a very
small fraction of a second before the dis-

charge. The induction might have been,
and probably was, taking place along many
paths. When, however, the insulator broke
down at the weakest point, and the spark
began to pass, the whole of the induction
at once transferred itself to the line of
discharge as being the path offering least
resistance, and then the breaking down and
relief of the strain was completed along
that path.

Given, then, that induction precedes

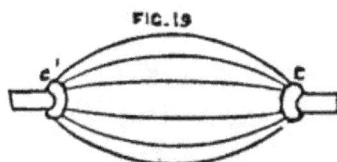

FIC. 19

discharge, if we see curved discharges, we
shall know that there was previously curved
induction. Let us look at the discharge of
a Holtz machine. It consists of a barrel-
shaped bundle of sparks. (Fig. 19.) Here,
in fact, are the curved lines of force,
or lines of induction, or lines of strain,

produced in visible shape before you. The
centre lines are straight, and the strongest
induction takes place along them; but induction strong enough to produce discharge
takes place in curved lines through all the
particles on all sides of the centre line.

These lines of force are real, and, I may
almost say, tangible things. They can be
attracted and shaped by the hand and other
conductors. I place my knuckle near the
lines, and they bend out towards it. This
means that the positively electrified particles of air induce negative electricity on
my hand, and then the two electricities
attract each other, and displace the whole
line of force. It would be difficult to con-
ceive the possibility of attracting an " action
at a distance."

We have shown that the induction is a
state of strain, and we have studied the
direction of the strain. We now ask, What is
its nature? Faraday showed experimentally

that the lines of force attracted each other, so that, if a number of them were side by side forming a " bundle," their mutual attraction drew them together as if the bundle had been tied up more tightly.

Maxwell has since pointed out that this is what occurs whenever a rope is mechanically stretched. The pull tending to lengthen the rope is accompanied by a pressure tending to make the rope thinner. To show you this *lateral* pressure, I have

FIC.20

here an india-rubber tube. When I stretch it the sides press on whatever is inside it, as you see. Whenever a mechanical tension occurs it is accompanied by a pressure at right angles to it.

Maxwell has shown mathematically that an electric induction acting as a tension along the lines of force is always accompanied by an exactly equal pressure at right angles to them. Electric induction, or tension, is a tension of exactly the same kind as the tension of a rope, and the medium which can support a certain induction force before breaking and allowing a spark to pass may be said to have a certain strength in exactly the same sense as a rope may be said to have a certain strength. Sir William Thomson has found that the electric strength of air at ordinary temperature and pressure is 9600 grains per square foot.

Finally, Mr. De la Rue has actually seen in one of his vacuum tubes a star of light showing a rain of particles thrown off at right angles to the main discharge.

LECTURE III.

Jan. 30.

Specific Inductive Capacity.

In the two previous lectures we have seen that Induction is transmitted from particle to particle of dielectrics, and that its phenomena are exhibitions not of some direct action passing through the insulator, but of something actually existing in the particles of the insulator itself; that it is in some peculiar straining of these particles that the causes of the phenomena will be found.

One of the first questions which now presents itself is, Do all insulators on which a given inducing charge acts suffer an equal strain, and therefore exhibit the same quantity of inductive action at the

E

other side, or, on the contrary, does the
same charge of electricity strain different
insulators differently and produce induced
changes of different strengths ; in other
words, are there in different insulators different capacities of receiving strain from a
given straining charge — differences of
specific "strainability"—that is, differences
of *Specific Inductive Capacity ?*

Various experiments, some of which I
shall hope to explain to you, satisfied
Faraday that the latter is the case, and
that different bodies have different specific
inductive capacities.

First, however, let us make it quite clear
what is meant by the term specific inductive capacity. Let a certain charge of
electricity be acting inductively across air
upon a neighbouring conductor, and let the
sizes of the conductors and the distance between them be such that the strength of the
charge induced on the second conductor is

equal to unity. Let the whole space between the conductors be now filled with some other insulator. The strength of the induced charge will now be no longer unity, but it will have some other value. The number which represents this value is called the "specific inductive capacity" of the substance between the conductors; in other words, the specific inductive capacity of a substance is the ratio of the inductive action across it to that across air. Air being taken as the standard, its specific inductive capacity is called unity.

We will now examine some of the various methods by which, from time to time, students of Nature have endeavoured to measure the specific inductive capacities of various bodies.

Here, at the risk of being tedious, it will be necessary for me to go somewhat into details of experiments; but it will, perhaps, not be entirely without interest

for us to look a little behind the scenes of
the laboratory, and see the kind of difficulties which beset the inquirer the moment
when, instead of, for instance, saying to
himself, as a mathematician can, "Let there
be a perfectly insulated charge of electricity
of given strength, at a given distance from
a conductor," he has to prepare himself to
say to his instrument-maker, " Here are
working drawings of an instrument which
is intended to place a given charge of
electricity in a given place; this portion
of the instrument is to regulate the charge,
that portion to measure it, and this other
portion to measure to 1-1000th inch its
distance from that conductor.

Can we construct this, so that every part
shall do its work, and no two parts shall
interfere with each other? Can we support it,
so that it will not shake, protect it from dust,
and yet contrive that neither the supports,
nor the cover, interfere with the induction ?

As Faraday was the discoverer of specific inductive capacity, we will begin with his experiments, and, through the kindness of Professor Tyndall, I can show you the very apparatus he worked with. Here it is (Fig. 21). Faraday's wish was to construct a Leyden jar, of which the metallic coatings should be fixed, and always in the same relative positions, while the insulator should be movable; so that various Leyden jars could be set up, which should be exactly alike in all respects, except in the nature of their insulator, which could be made to consist either of air, glass, sulphur, or any other substance.

If such jars could be constructed, and if differences were observed in their behaviour, these differences could only be due to differences of induction through the different insulators, or to differences of specific inductive capacity.

The apparatus consists, as you see, of a

metal ball, which can be surrounded by a
larger hollow one.   The outer ball is made
in two pieces, so as to allow the inner one
to be placed inside it.   There is a space of

FIG. 21.

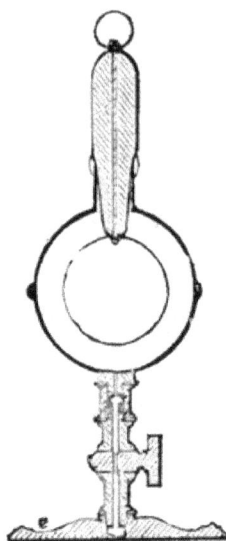

0·62 inch between the surfaces of the balls.
The inner ball is supported by an insulating
stem of shellac passing through a hole in
the outer one.   A wire which passes
up inside the shellac allows the inner

ball to be put in connection with an electric machine, electrometer, or other apparatus.

The space between the balls contains the insulator. It may be air, as at present, or the whole or part of the space may be filled with glass, sulphur, &c. Faraday preferred only to fill half the space, and then to calculate what the effect would have been if the whole space had been filled. He, therefore, prepared his insulators in the form of hemispherical cups. Here are some of them.

He constructed two of these Leyden jars, so that he could observe simultaneously their actions with different insulators, and endeavoured to make them precisely alike. If they had not been precisely alike, there would have been a difference in their behaviour which would have been due, not to difference in the specific inductive capacities of the insulators, but to differences in the shape and size of the jars. In order to make sure that they were exactly alike, he made an elaborate series of preliminary experiments. We need not go into all the details of these preliminary experiments, but we can indicate the principle of them in a few words.

The object of them is, we remember, to determine whether the two machines have equal capacities for electricity. That is, whether under similar circumstances they will each hold an equal quantity of electricity.

To determine this, Faraday first charged one apparatus only, and measured the charge. He then connected the two machines together, so that the charge divided itself between them. He then separated them and re-measured the charge of the first one. If the second apparatus had the same capacity as the first, it would have taken away exactly half the charge.

If it had a greater or less capacity, it would have taken more or less than half the charge.

The machines were, therefore, adjusted till the charge left in each after division with the second was exactly one-half of the original charge before division. I have said that " the amount of charge was measured," but have not yet explained how that was done. Here again as the electro*scopes* which we have been experimenting with only show the existence or non-existence of a charge, but do not measure its amount, we require an electro*meter*. The electrometer used by Faraday to measure the induction was the invention of Coulomb, and is called the " torsion balance." Descriptions of it will be found in all books on physics. The numbers which Faraday gives as the relative strengths of electrifications are the number of degrees of twist which, for these electrifications, he read off on the torsion balance.

Having adjusted his torsion balance and determined the equality of his two Leyden jars, Faraday was ready to commence his measurements of specific inductive capacity. He kept one apparatus full of air, and placed in the other this hemispherical cup of shellac. He then compared the inductive actions through the two machines, and he at once found that the induction through the shellac apparatus was greater than that through the air apparatus in the proportion of 176 to 113, or 1·55 to 1. In this case the air apparatus had been charged first.

Another set of experiments in which the lac apparatus was charged first gave a ratio of 1·37 to 1. This difference, which is considerable, is accounted for by the fact that the experiment takes some time, and that there is a constant leakage of electricity going on; and in the one set the effect of the leakage would be to give too high a result, and in the other to give too low a one. The mean, therefore, will not be far from the truth.

Faraday gives as his result that the induction through the apparatus half-filled with shellac is 1·5 times that through the one full of air. From this he calculates that the ratio of the specific inductive capacity of shellac is to that of air rather more than 2 to 1.

I have purposely avoided attempting to give the exact details of Faraday's method of working, for two reasons. One is that it is an exceedingly difficult thing to understand, as the inductive actions through the different insulators are compared by an indirect method, to follow which requires a considerable familiarity with the laws of induction; and the other is that I thought it unnecessary to burden your memories with the minor details of a method of working, which, owing to the invention of improved apparatus, no experimenter would now adopt, wonderful as it was in its day, and wonderful for all time as *are* the results which were obtained by it.

Faraday continued his experiments on other substances, and here is a general table of his results : Shellac, 2 ; sulphur, 2.24 ; flint glass, 1·76 or more.

After Faraday came numerous experimenters, who have published results more or less accordant for the specific inductive capacities of many insulators. As time will not admit of my giving you an account of all the methods which have been used, and as for three years I myself have been engaged in a determination of the specific inductive capacities of various substances, I have preferred to give you an account of my own experiments only; partly because I believe they are the latest which have been made, but more particularly as I shall be able to make a more interesting lecture about methods with which I am practically familiar, and which have, so to speak, grown up under my hand, than I could if I told you of methods which I have only read about.

First, however, let me tell you why it is so important that we should have accurate knowledge of the specific inductive capacities of the various substances whose names you see in the table (page 90), india-rubber, vulcanite, paraffin-wax, glass, gutta-percha, &c.

One reason is that these substances are much used for the insulating parts of electrical and telegraphic instruments, and unless we know their specific inductive capacity we cannot tell beforehand what their effect will be in any particular instrument.

Also, some of these substances are used

for the insulators of submarine cables.
Now, the speed of signalling, and with it,
of course, the gross receipts of the telegraph company, depends, among other
things, on the specific inductive capacity of
the insulator, so it is as well to know
this accurately before manufacturing the
cable. The speed of signalling depends
perhaps more upon the specific inductive
capacity of the insulator than on anything
else. The lower the specific inductive
capacity the greater the speed. The great
object of telegraph engineers at present is
to discover a good insulator of very low
specific inductive capacity.

But there is another and to us a far more
important reason for desiring accurate
knowledge on this point.

Every new investigation which is made
points to a close connection between
electricity and light. The theory of their
connection, which I shall hope to explain

in the next lecture, requires a certain relation between the specific inductive capacities and certain optical properties of transparent bodies. This theory, which may even tell us what electricity is, can only be tested by an accurate knowledge of the specific inductive capacities of transparent bodies.

The experiments which I am going to tell you about have been carried on under Professor Clerk Maxwell's advice and superintendence, and he is the inventor of the new method which has been used.

The great difficulty which all previous experimenters have met with was to make the experiments quickly enough. If the electrification is allowed to continue for even 1-100th of a second all sorts of effects come in which are not due only to the specific inductive capacity of the insulator.

Three required conditions were before us when the new apparatus was being designed.

1st. The experiments must be made very quickly. This was accomplished by reversing the electrification 12,000 times per second, so that the duration of each charging of the insulator was only 1-12000th of a second.

2nd. The electrified metal plates must not touch the insulators. When they are allowed to do so it is found that there is a leakage through the insulator which affects the results.

3rd. The method must be a zero method. That is, instead of measuring the strength of two inductions separately, and then comparing them with each other, we must balance the two inductions against each other until the result is zero.

As an illustration of the two methods of working, let us set to work to determine the weight of a piece of butter. We ordinarily use a zero method; that is, we put the butter in one pan of a scale, and vary the weights in the other, till there is *no deflection* of the beam. A non-zero method, such as were the early determinations of specific inductive capacity, would have been to have put the butter in one pan, a fixed weight in the other, and to have endeavoured to calculate their ratio from the deflection of the beam.

In the experiments which I am about to describe, the induction through a given thickness of the substance under examination is opposed to the induction through a thickness of air which can be varied till the two actions exactly balance. A comparison of the thickness of the air and the substance gives the specific inductive capacity of the latter. The same electrification being used for the two actions accidental variations in it do not affect the result.

As the apparatus before you is exceedingly complicated, we shall do better to

F

first study this diagram (Fig. 22), where only
the most essential parts are shown. We may
here mention that substances across which
induction takes place are called " dielectrics," from the Greek preposition διά,
across.

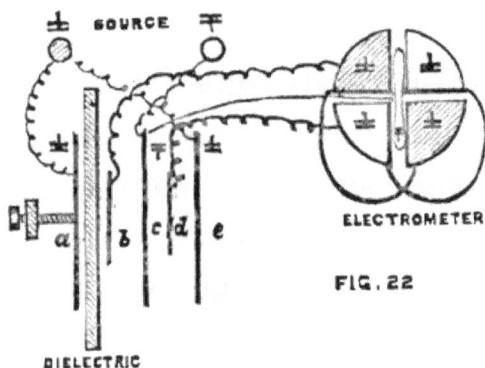

first study this diagram (Fig. 22), where only

FIG. 22

The induction balance consists of five
circular metal plates, a b c d e, seen edgeways in Fig. 22, fixed and insulated parallel
to each other. b c d e are about an inch
apart. The distance from a to b can be
varied, by means of a screw, from about
two and a half inches to nothing. a c e
are six inches diameter, b d four inches.

a and *e* are connected to one pole of the source of electricity, *c* to the other. *b* and *d* are connected to the quadrants of a Thomson (Elliott pattern) electrometer.*

At the top of the diagram are shown the two poles of the source of electricity, which are always oppositely electrified, but are constantly being reversed. The dielectric under examination can be inserted or removed at pleasure between *a* and *b*. The centre plate *c* is also connected to the needle of the electrometer.

Let us now suppose that the dielectric is removed, and that all the plates are arranged symmetrically, viz., that distance *c b* equals distance *c d*, and distance *e d* equals distance *a b*. Suppose for a moment the reversing apparatus stopped, and the electrification to be that of the upper signs in the diagram. Let us examine the effect on the electrometer.

* See pp. 35, 36, and foot-note.

The inductive actions of c, on b and d, are equal and similar, consequently, the effect of c on the electrometer is zero, for all four quadrants are equally and similarly electrified by it. The inductive action of a on b is equal and similar that of e on d, and consequently, the effect of a and e on the electrometer is also zero, and thus, however strongly the plates are electrified, there will be no deflection of the electrometer as long as the arrangement is symmetrical.

Now, however, let the dielectric be introduced between a and b. By reason of its specific inductive capacity being greater than that of air, the action of a on b, which passes through it, will be greater than that of e on d, and consequently, though all four quadrants are still similarly electrified, the electrification of the shaded quadrants will be strongest, and the needle will be deflected.

Let, now, the screw be worked, and plate a moved so as to increase the distance $a\ b$. The action of a on b will be diminished, and when a has been moved so far that the needle has again come to zero, we shall know that the increase of the distance between a and b which has been made by moving a has diminished the induction by an amount exactly equal to the amount by which it was increased by the greater action through the dielectric under examination.

Knowing the thickness of the dielectric, and the amount which a has had to be moved, we can calculate its specific inductive capacity.*

---

* The formula of calculation used is as follows : Let us write K specific inductive capacity, b thickness of dielectric, a_1 reading of plate a with only air in the balance, a_2 reading when dielectric is inserted. The action across a dielectric of thickness b and specific inductive capacity K is the same as that across a thickness of air $\dfrac{b}{K}$. Hence, when we introduce the dielectric,

Let us now set the reversing apparatus
to work, and suppose the equilibrium not
to be established. Suppose that we have
inserted our dielectric but have not moved a.

At first let the direction of the electrification be that of the upper signs (Fig.
22). Then there will be a deflection
of the electrometer needle in the direction

we do the same as if we had put in a stratum of air
thickness $\frac{b}{K}$, and, at the same time, take away the
stratum of air thickness b, which is displaced by it.
Hence, the insertion of the dielectric is the same as
if we had decreased the distance between a and b
by a quantity equal to $b - \frac{b}{K}$. But to produce an
equal or opposite electrical effect—$i.e.$ to bring the
needle to zero—we increased the distance by an amount
$a_2 - a_1$. Hence these two quantities are numerically
equal, and we have $a_2 - a_1 = b - \frac{b}{K}$; that is,

$$\frac{K}{b} = \frac{1}{b - (a_2 - a_1)}, \text{ or } K = \frac{b}{b - (a_2 - a_1)}.$$

opposite to the motion of the hands of a watch, that is, to the left. Now let the electrification be reversed. If the needle were charged in the ordinary way, and remained positive, there would be a deflection to the right. When the reversals were rapid the alternate impulses to right and left would neutralize each other, and there would be no deflection, however much the equilibrium was disturbed.

To get out of this difficulty a plan was designed which, as it is due to Professor Maxwell and not to myself, I may call beautifully ingenious. Instead of keeping the needle permanently charged it is connected to the centre plate c and reverses with it.

First, let us suppose the electrifications have the upper signs, and that by introducing the dielectric we have made the shaded quadrants the strongest. The force will be attractive, and the needle will turn to the left. On reversing the electrifications so that they are all those of the lower signs, the shaded quadrants will still be the strongest, the force on the needle will still be attractive, and it will still turn to the left.

In practical work, when the electrifications of the five plates, the dielectric, the four quadrants, and the needle, are all being reversed 12,000 times per second, the needle is perfectly steady, and so exactly under the control of the screw of a, that a motion of a of 1-1000th of an inch nearer to or further from b produces a perfectly visible motion of the light-spot which indicates the motions of the needle.

We now turn from the mathematical diagram to the actual instrument. Here you see we are hampered by many troublesome conditions; we have not only to say, " Let a b c d e be plates in such a posi-

tion," but to support them in that position
without the supports interfering with their
action; not only to say, "Let them be
insulated," but to make sure that no
electricity escapes; and again we have not
only to say, "Let the dielectric be removed and replaced," but to provide means
for removing it. You see we must not
put our fingers in between the plates when
the apparatus is in action.

Well, here is the apparatus (Fig. 23), as
it grew up in the course of three or four
months' labour. Four of the plates,
viz., *b c d e*, are supported from brass
stages, carried on brass pillars. Each
plate hangs by a thin steel rod, which is
rigid enough to move the plate a little out
of the vertical, if required for adjustment.

The upper end of each steel rod passes
through a hole in the stage, and is attached
to the centre of a little ebonite triangle e,
which rests on the stage by three levelling

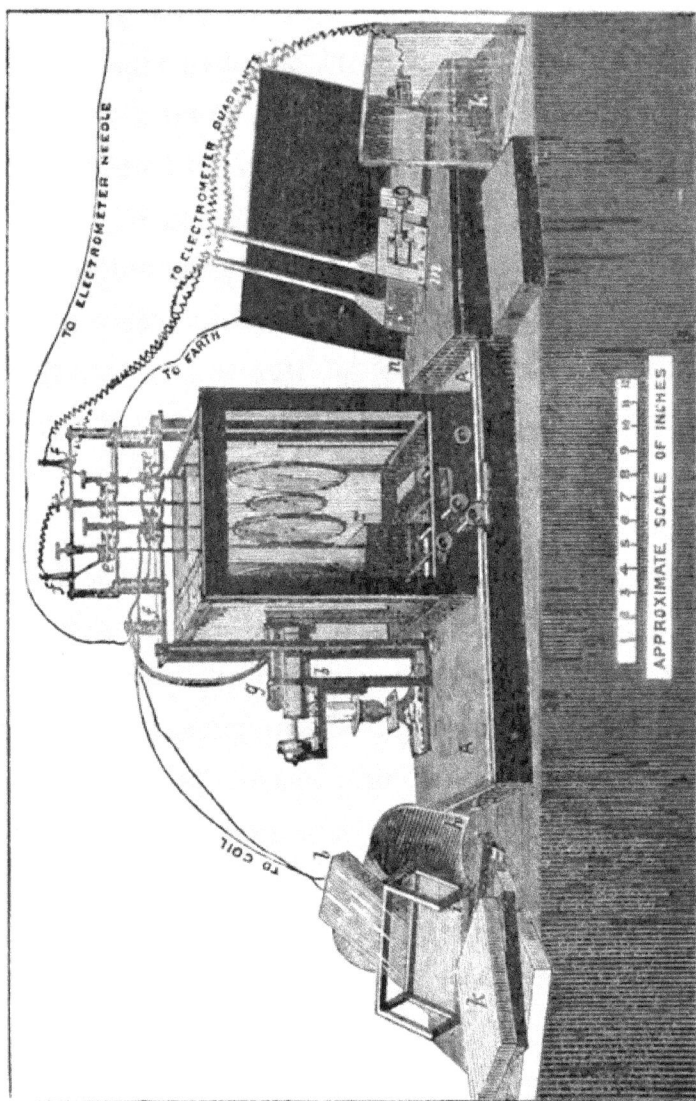

TO ELECTROMETER NEEDLE

TO ELECTROMETER QUADR

TO EARTH

TO COIL

APPROXIMATE SCALE OF INCHES

FIG. 23.

screws at its angles. By means of these screws, and by turning the steel rods, the plates can be adjusted so as to be exactly parallel with each other. When the plates are adjusted, they are clamped by screws, which come down upon each triangle from the stage above. Wires connected to the steel rods lead from each triangle to the binding screws f f, by which wires from other instruments can be attached to them.

The remaining plate a is supported differently. An ebonite block is fixed at the back of it, and this is fixed to a brass rod b, about an inch thick, of the shape ▽, and about six inches long. This slides in two V-shaped grooves in two brass pillars b, and is pressed down into them by springs. It is moved by a very delicate screw, with a specially contrived spring collar, to avoid " back-lash," as it is called ; that is, to insure that the motion of the plate reverses at the same instant as the motion of the screw-head is reversed.

A scale is engraved on the sliding rod, and a vernier on one of the supports. This scale is read to 1-1000th inch by a telescope fixed on the case of the instrument behind g (it cannot be seen in the engraving). The scale is illuminated by a candle.

The dielectrics k k are made seven inches square, and their thickness varies from one-quarter inch to one inch.

The slide for inserting and withdrawing the dielectrics is shown, c d, in the picture. It moves in guides, and can be pulled in and out by the square handle without opening the case, or disturbing the experiment in any way. The three other handles are for adjusting the position of the dielectric; one moves it parallel to itself by means of a rack, one turns it round a vertical axis by a tangent screw, and the third round a vertical axis by screwing in or withdrawing a wedge under one side.

The callipers m shown in the same figure are for measuring the thickness of the dielectrics, and for adjusting the plates parallel to each other. For this latter purpose they are laid on the hinged bracket, n, which can be fixed to a socket at the back of the apparatus.

In work all the upper stage which contains the connections of the plates receiving the induction is enclosed in a metal box connected to earth, the wires leading to the electrometer are enclosed in a metal tube, and the electrometer itself is in a metal case. The reason of this is to protect these parts of the apparatus from accidental induction from the connecting wires, &c. If from any accidental cause the earth connection is interrupted, the effect is at once seen in the uncertain and irregular behaviour of the instrument.

The metal cover, h, first used for the upper stage, is shown. It was made of card and tinfoil. A brass one is, however, now used instead of it.

Before going any further I must express my obligations to Mr. Kieser, of the firm of Elliott Brothers, for the admirable way in which he has constructed the instrument from my drawings.

The electrometer is of the ordinary Elliott pattern quadrant described by me in the last lecture in the account of Hopkinson's experiment.

The source of electricity is a large induction coil by Apps, having twenty-two miles of secondary wire (Fig. 24), which, with a suitable battery and break, is capable of giving a seventeenth-inch spark in the air. In these experiments it is, however, used in a different manner. The object is to obtain a moderate electrification, very rapidly reversed, and to

insure that the positive and negative electrifications shall have equal strengths.

When a coil is worked in the ordinary
way, although the same quantities of elec-

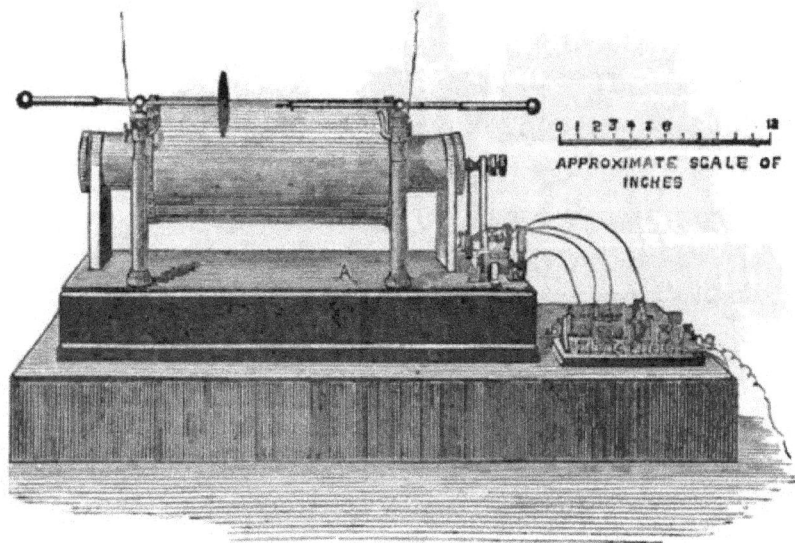

FIG. 24.

tricity are produced both on making and
breaking the primary, yet the arrangement
of the currents is such that the current
produced in one direction on breaking will
produce a much stronger external effect

than that produced in the other direction
on making. To obtain equal electrifications
in the two directions it is necessary to use
a very large coil, a very small battery, and
a very rapid break.

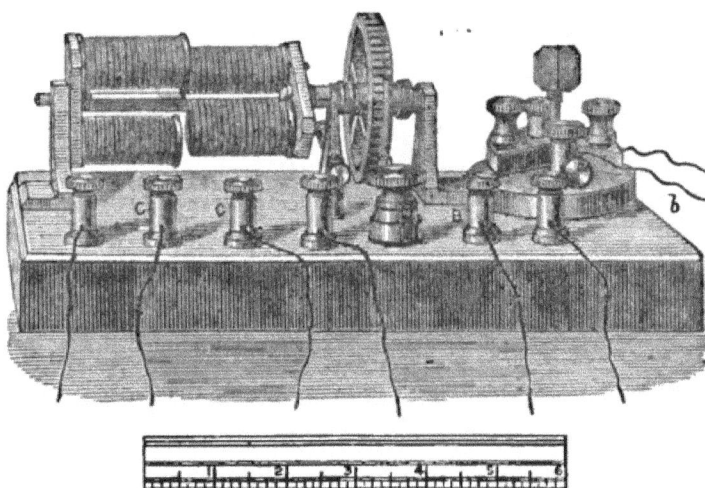

APPROXIMATE SCALE OF INCHES.

FIG. 25.

In these experiments the current in the
coil primary is only that of ten small
Leclanché cells. The rapid break is shown
in Fig. 24, and also on a larger scale in
Fig. 25. It consists of a little electro-mag-

netic engine. The scale of inches in Fig. 25 shows the size. One electro-magnet is fixed, and another revolves. A commutator is so arranged that the force between approaching poles is always attractive, and that between poles which are moving apart repulsive. When the engine is worked by four quart-sized Grove cells the flywheel revolves just 100 times per second. You see the whole engine is not more than eight inches long and four high and broad, yet when I set it in motion the hum and vibration are felt all over the building.

In the rim of the flywheel, which is about two inches diameter, are 60 slits cut, into each of which is let a piece of ebonite. A light spring presses on it, and the primary current on its way from the Leclanché battery to the coil has to pass from the spring to the wheel. It is thus broken sixty times, and closed sixty times in each revolution of the wheel. At each "break"

G

a current is induced in one direction in the secondary, and at each "make" one is induced in the other direction. Thus there are 120 alternating currents induced every revolution of the engine, and, as the engine turns 100 times per second, there are 12,000 currents each second. The engine does great credit to its maker, Mr. Apps.

To test the equality of the currents in the two directions, the secondary poles were connected to a small vacuum tube. No effect whatever was produced on the light by reversing the primary by means of a commutator between the engine and the Leclanché battery.

One of the secondary poles of the coil is connected to plate c of the induction balance and the other to the plates a and e, and of course the electrifications of every part of the balance and electrometer are reversed with those of the coil poles. The strength of the electrification is such that a spark

could be obtained between the coil poles of
from 1-25th to 1-50th of an inch; that is,
it is about equal to the direct discharge of
2000 silver cells.

In case, however, the sum of the make

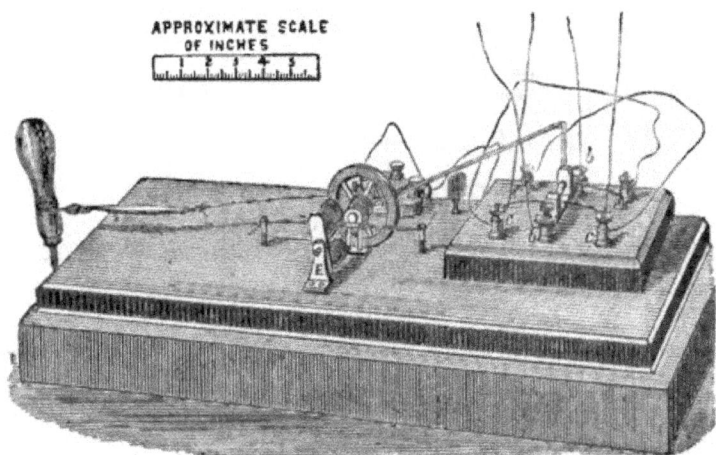

FIG. 26.

and break electricities should not be exactly
equal, the machine (Fig. 26) was introduced
for reversing the secondary current about
thirty times per second on its way from the
coil to the induction balance.

It consists of a wire frame, which dips

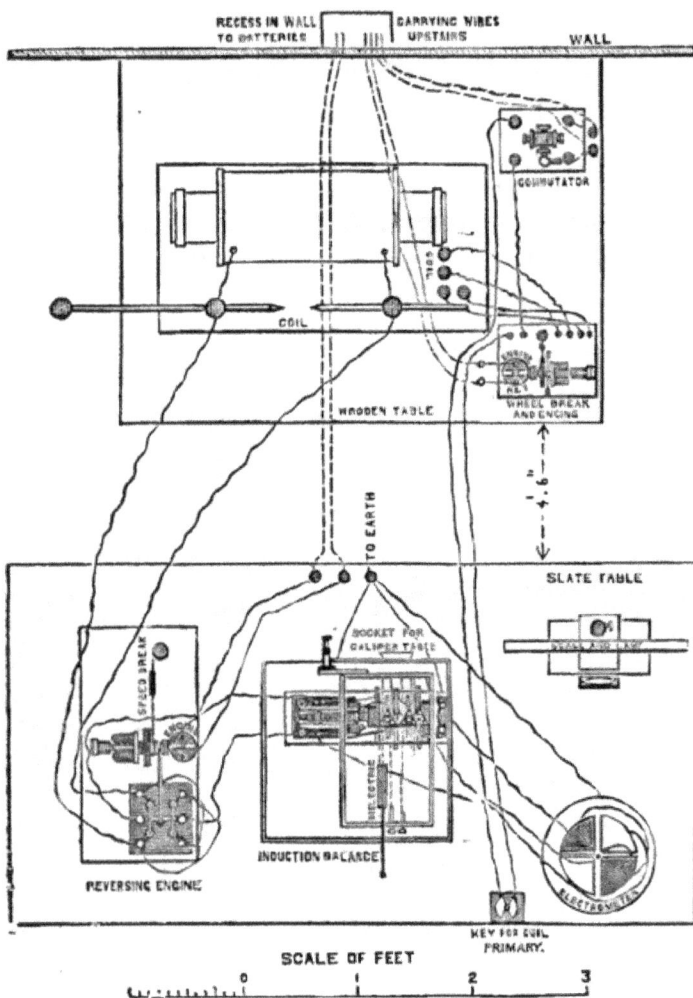

FIG. 27.

alternately into different mercury cups, and so reverses the current. It is worked from a crank on the axis of a small engine, similar to that of Fig. 25. The speed is regulated by a friction brake, consisting of a silk loop round a pulley on the axle, to which an india-rubber band is attached; a cord from this band is wound round a bradawl, driven into the base; by turning the bradawl, the speed can be exactly regulated. This engine is, however, not now used, as the reversals given by the rapid break are, I believe, perfect.

This plan (Fig. 27) shows the arrangement of the apparatus in my laboratory. You see that, standing where I can see the scale, I have the key of the coil primary under my right hand, and the mechanical slide and the screw of a under my left. Compare this diagram with Fig. 22.

Here in the lecture-room the apparatus is arranged in nearly the same way, except that, instead of the small scale and paraffin lamp, we have a large scale and lime-light, so that you may see the deflections of the electrometer, and that Mr. Cottrell has hinged a long wooden pointer to the induction balance, so that you can all see the motions of plate a.

We will now make an actual determination of the specific inductive capacity of this beautiful slab of " double extra dense flint " glass. I place it on the slide and draw out the latter, so that the glass is not between the plates, and now put plate a into what I think will be about the position of equilibrium. We start the engine and make contact in the primary circuit, and you see there is a small deflection. I move a backward and forward till the light-spot comes to zero and remains there as we make and break the primary. We now read the scale of a and write down the result as a_1.

Now I push the handle of the slide and

insert the dielectric between a and b. You see that there is at once a large deflection to the left. I work the screw of a, and draw it further away from b, and here you see is the light-spot coming back to zero, and now, when I have moved a about an inch, I have adjusted the electrometer exactly to zero. We take the reading again and write it down as a_2.* The difference between the two readings is the motion of a, and knowing this and the thickness of the glass, we can calculate its specific inductive capacity by the formula given in the foot-note to page 69.

* In practice the probable position of a_2 is always obtained by a preliminary experiment, and the plate a put there before making contact with the dielectric inserted, so that there is never any large deflection. It is found that when large deflections take place, the needle does not always return to quite the same zero, In the experiment shown to the audience, however. this was not done in order that the disturbance of equilibrium caused by inserting the dielectric might be more clearly seen.

On drawing out the glass, there is, of course, a deflection to the right, but you notice it happens, from some defect in the adjustment of the electrometer, that this deflection is very much smaller than was the previous deflection to the left.

Here, however, we may see the beauty of the zero method, for we have nothing to do with the magnitude of the deflection, and we see that if we bring the needle back to zero by screwing in a, we require exactly as much inward motion to compensate this small deflection to the right as we previously required outward motion to compensate the large deflection to the left. In fact, when the light-spot is again at zero, the reading of a is precisely the same as it was at the beginning of the experiment.

In actual work, successive readings never disagree by more than two, or at most three thousandths of an inch. We cannot, however, get the same accuracy

here, as the rapid break, and the audience
are all supported on the same floor as the
induction balance and electrometer, and so
cause vibrations. In my own laboratory
the induction balance and electrometer are
supported on brick and slate piers quite
independent of the floor on which the break-
engine and the observer stand. Here on
this diagram (page 90) are the results of
my experiments, extracted from my paper
which has recently been read before the
Royal Society.*

If we look at the experiments on ebonite,
we shall see that we get substantially the
same results with dielectric plates of very
different thicknesses. This is a very good
test of the accuracy both of the instrument
and the formula of calculation. The experiments on paraffin wax agree very well
with each other. The most accurate determinations, which have been recently made

* Proc. Roy. Soc., 191, 1878, p. 155.

Dielectric.		Specific Inductive Capacity.
Glass, Slabs about 1 inch thick. Chance's optical glass.	{ Double extra dense flint . Extra dense flint . . . Light flint . . . . . Hard crown	3·1639 3·0536 3·0129 3·1079
Common plate, two slabs.	{ No. 1 3·2581 { No. 2 3·2282 }	. . 3·2431
Ebonite, four slabs, $\frac{3}{4}$, $\frac{1}{2}$, $\frac{1}{2}$, $\frac{1}{4}$ inch.	No. 1 2·2697 No. 2 2·2482 No. 3 2·3097 No. 4 2·3077	. . 2·2838
Best quality gutta-percha	. . . .	2·4625
Chatterton's compound	. . . .	2·5474
India-rubber . .	{ black . . . . . . { vulcanized	2.2200 2·4969
Solid paraffin, sp. gr. at 11°C. ·9109. Melting point 68° C. Six slabs cut in planing machine. Results corrected for cavities.	No. 1 1·9940 No. 2 1·9784 No. 3 1·9969 No. 4 2·0126 No. 5 1·9654 No. 6 2·0143	Mean 1·9936
Shellac . . .	. . . .	2·7464
Sulphur . . .	. . . .	2·5793
Bisulphide of carbon .	. . . .	* 1·8096

---

* I am not quite certain of the accuracy of this result.

of specific inductive capacity, have been those of Messrs. Gibson and Barclay,* who experimented on paraffin only. They used a method entirely different from mine, and found that the specific inductive capacity of their paraffin was 1·977. Correcting for a slight difference of density, I find that if they had used my paraffin their result would have been 1·9833, which differs from my result by only one-half per cent., or one part in two hundred.

* Phil. Trans., 1871, p. 573.

LECTURE IV.

SPECIFIC INDUCTIVE CAPACITY OF GASES.
ELECTRO-MAGNETIC THEORY OF LIGHT.

FARADAY* made a great number of experiments on the specific inductive capacities of gases, to see if he could detect any differences between them, but he was unable to do so. He compared no less than twenty-five pairs of gases. He also compared dry and damp air, and hot and cold air, and air at various pressures, but with his apparatus he could detect no difference at all.

It is curious to see how long and how earnestly he sought for evidence of differences of action.

* " Experimental Researches," vol. i. p. 406.

He seemed to have so strong an instinct that there ought to be a difference, that he literally struggled against the evidence that every experiment he made seemed to pile up against his theory. He knew that gases differed in so many of their other physical properties that he could hardly believe they were all alike in this one.   He used the same apparatus as I showed you last lecture. It was not till 1877 that Professors Ayrton and Perry,* working with apparatus many thousand times more delicate than that which was at Faraday's disposal, succeeded in showing that the reason why Faraday had not been able to detect differences of specific inductive capacities in gases was not that these differences did not exist, but that they

---

* " On the Specific Inductive Capacity of Gases." Paper read before the Asiatic Society of Japan, April 18, 1877. Printed at the *Japan Mail* Office, Yokohama.

were too small to be detected, except by a
quadrant electrometer. Professors Ayrton
and Perry have not only shown that
different gases have different specific
inductive capacities, but that the specific
inductive capacity of the same gas is
different at different temperatures and

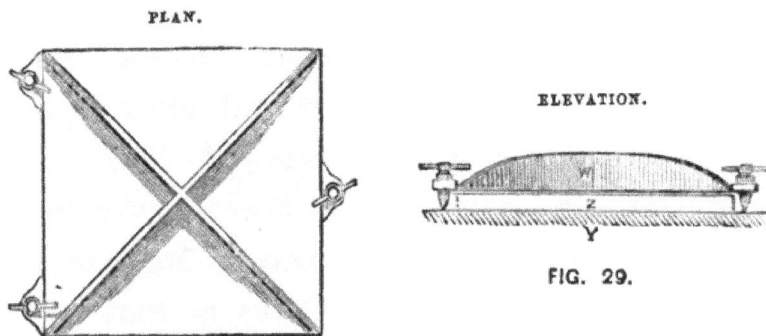

PLAN.

FIG. 28.

ELEVATION.

FIG. 29.

pressures ; and, further, they have actually
measured the amounts of these differences.
Through the kindness of Professor Ayrton
I am able to explain to you his method of
working ; but I cannot show you his
experiments, as they were made in Japan,
and the apparatus being the property of

the Japanese Government, had to be left behind when Professor Ayrton returned last August.

Professor Ayrton's method of working was as follows:—He prepared two condensers—that is, Leyden jars—into one of which he could put different gases as the insulator; one, which he called the "open air condenser" (Figs. 28, 29), consisted of a thick brass plate, Z, laid on the table, Y, having over it another brass plate, W. This latter was supported on three ebonite levelling screws. The two brass plates formed the conductors, and the air between them formed the insulator. You will see the use of this condenser immediately.

The other condenser was called the "closed condenser." (Figs. 30, 31, 32). It consisted of eleven brass plates, fixed parallel to each other, in a metal box. Nos. 1, 3, 5, 7, 9, 11 were all connected to one piece of metal, and Nos. 2, 4, 6, 8, 10 to another. The

two sets of plates were insulated from each
other. They thus formed a condenser of
very large surface, and all the spaces

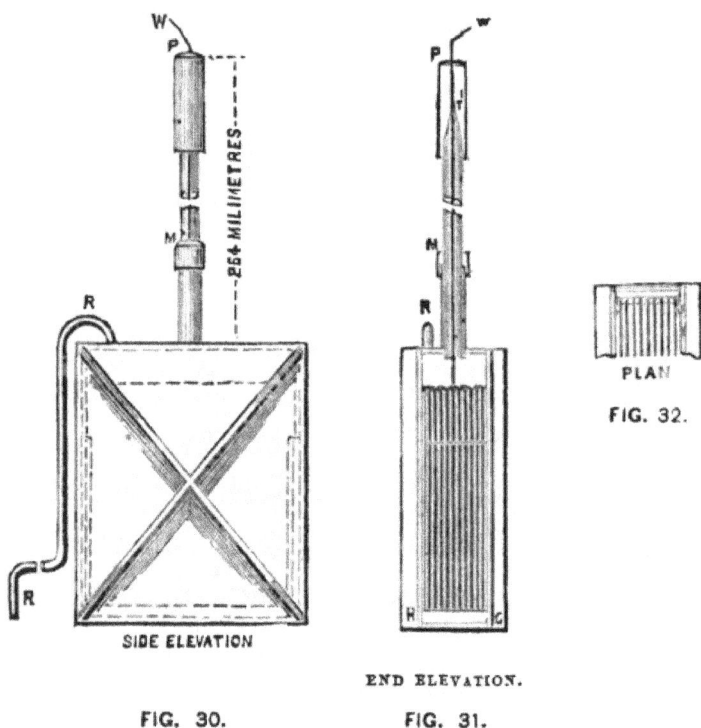

SIDE ELEVATION

FIG. 30.

END ELEVATION.

FIG. 31.

PLAN

FIG. 32.

between could be filled with the gas under
examination. The brass box was closed
quite air-tight, and could be connected to

an air-pump, by the tube R, when required. The induction through this apparatus, when it contained different gases, could be compared. But it would be impossible to preserve a perfect record of the action through the condenser, and to insure that the inducing force should always be the same. For this reason the " open condenser " before mentioned was used.

The closed condenser was filled with dry air at ordinary pressure, and the induction through it was compared by means of a very delicate quadrant electrometer with that through the open condenser. This ratio was noted.

The closed condenser was then filled with some other gas, and the ratio again noted between it and the open condenser. The open condenser, of course, remained the same. The ratio of these two results gave the ratio of the inductive actions through air and the gas under examination.

H

The open condenser, in fact, only acted as a standard measure, just as if it were desired to compare the lengths of two ropes, and it was not convenient to lay them together, they could each separately be compared with a yard measure.

The difficulties of this investigation must have been enormous.

Professor Ayrton's paper contains quite a heartrending list of breakages, leakages, twistings. He relates how, first, the box was not air-tight. Then a smith, having been sent for to solder it, his hot tools damaged the ebonite inside, and it all had to be taken to pieces again. Then the long glass tube, P M, in which the conducting wire, W, was insulated broke again and again. Then, when all was finished, mercury from the air-pump got inside, and spoilt the whole affair.

We in England suffer in the same way; but can we imagine what the difficulties of delicate physical investigation must be when all repairs of instruments have to be done either by the experimenter himself, or by Japanese workmen?

However, at last the skill and invincible patience of the investigators conquered all difficulties, and the results on this diagram were obtained :—

AYRTON AND PERRY ON SPECIFIC INDUCTIVE CAPACITIES OF GASES.

Dielectric.	Specific Inductive Capacity.	Dielectric.	Specific Inductive Capacity.
Air	1·0000	Hydrogen	0·9998
Vacuum	0·9985	Coal Gas	1·0004
Carbonic dioxide	1·0008	Sulphuric dioxide	1·0037

These results are a marvel of experimental work.

In some cases the whole difference which had to be measured was only two parts in 10,000. The maximum difference was only thirty-seven in 10,000, and yet these differences have not only been observed, but have been measured so accurately that a mathematical comparison of the various observations of which they are the means shows that the probable error is not more than five parts in 100,000, or 1-200th per cent.

In future it will not be sufficient to say, the specific inductive capacity of air is taken as unity. We must specify the pressure and temperature of the air, and say that the unit specific inductive capacity is that of dry air at 30 inches barometer and 32° Fahr. temperature.

We now come to perhaps the most important and interesting part of our subject; namely, the relations between electricity and light.

It is chiefly by the consideration of the connections which may exist between different forms of physical energy that we may hope to some day obtain a clearer notion of their actual nature.

Between light and electricity there are numerous and close relations and analogies. The form, however, in which electricity is most intimately connected with light, is not the static form, which we have been considering, but another, called " electro-dynamic," or, because of its magnetic properties, " electro-magnetic." Hitherto in these lectures we have considered only bodies charged with electricity at rest. We have examined the induction of charged bodies, that is, bodies containing electricity, but containing a certain quantity not in motion. The only cases of the motion or flow of electricity which we have noticed have been the momentary motions which have accompanied the discharging or charging of conductors.

When, however, electricity instead of being at rest, is *flowing* as a current—when, for instance, electricity is being drawn out of one end of a wire and constantly being

renewed at the other—it produces an entirely new set of inductive actions, different
from those which we have been considering.
Now I am not going to attempt, in the
half-hour which remains of the last of these
lectures, to give an account of the laws of
electro-magnetic induction ; but this much
will be necessary, and, I think, sufficient
as an introduction to our study of the
relations between electric induction and
light.

Electric (that is, electro-static) phenomena
are so closely linked in every detail with
electro-magnetic ones, that any arguments
which show that the same mechanism
transmits electro-magnetic induction and
light will also hold for electro-static
induction, though I do not say but that
the working of the machinery may be very
different in the two cases.

" One mechanism for electric induction
and light." This is Professor Clerk Max-

well's theory,* which every new measurement which is made helps to render more probable.

It is this theory which, when it is finally proved, and when some difficulties that now beset it are cleared away, as no doubt they will be, will tell us "what is light," "what is electricity." Let us consider it, and try and understand what it means.

I will give you the substance of the preface to Maxwell's chapter on the subject, and then explain to you such of his arguments as I understand well enough to put into an unmathematical form.

Light is, all men of science are now agreed, a wave motion of a medium which we call the ether, which fills all space, and probably permeates all bodies.

The sun expends energy, and sends it off from him. This energy travels through

* "Electricity and Magnetism," by J. Clerk Maxwell, F.R.S., chapter xx., vol. ii. p. 383.

the dark planetary spaces until it falls on
the eye, and then it is felt as light.   When
it falls on opaque bodies, part is reflected
from them, and, falling on the eye, is felt
as the light which makes them visible.
Part penetrates them, and heats them.
In what form did this energy which we
know as light and radiant heat exist in the
dark space between the sun and earth?

The undulatory theory of light answers
that this dark space is full of a medium, a
very thin fluid, and that the energy given
off from the sun is expended in producing
waves in that portion of the medium next
it, which in their turn expend the energy
they have received in producing waves in
the next portions; and so the energy is
transmitted by these wave motions, until
on striking the earth or the eye it becomes
heat or light.

We have said that electro-static induction
is a strain or distortion of the insulator

through which it is transmitted. This medium may be air, or glass, or paraffin, &c., as we saw. But what is the medium when the sun acts inductively on the earth, as no doubt he does?

When a sun-spot bursts out into stronger activity all the magnetic instruments at Kew Observatory move in sympathy with it.

What is the medium which transmits this electro-magnetic induction from the sun to the earth? Professor Maxwell says that it is one and the same medium as that which carries the waves of light, or that *light itself is an electro-magnetic disturbance.*

Let us now consider some of the arguments which have led up to this theory.

It is proved, I think, that electric induction is a strain of some kind; and, when electric induction passes through space in which there is not any ordinary matter, we agree to call the unknown

something that fills the space and transmits
the strain an "ether."

Light is a strain of some kind; and
when light passes through space where
there is not any ordinary matter, we agree
to call the unknown something that fills
the space and transmits the strain an
"ether."

One word of explanation of the term
"strain." In physics this word has a
more extended meaning than in common
language. Any change of form whatever
is called a strain. A wave motion would,
therefore, be called a strain.

How shall we decide whether these two
ethers are one and the same? We must
examine and measure as many of the
properties of each ether as we can; and
then, if we find that all the properties are
the same, we shall be sure that the ethers
are not two but one.

If, again, we find that most of their

properties nearly agree, but not quite, we must reserve our judgment; but we might in that case be allowed to speculate on the possibility of the same ether sea vibrating somewhat differently when disturbed by electricity or by light.

One important point of resemblance appears at once. In the case of light, the researches of Young, Fresnel, Huygens, and Green have shown that the energy in the medium is partly "potential" and partly "kinetic."

These are two hard words, but I think I can make their meaning clear to you.

"Kinetic" energy is the energy of motion. "Potential" energy is the energy of strain. A stone when falling has kinetic energy. By virtue of its motion it can strike a hard blow on anything that comes in its way and stops that motion.

The same stone suspended by a string has potential energy, because the instant

that the string is cut it will acquire kinetic
energy. We may regard the earth and
the stone together as a system which was
strained when the stone was pulled away
from the earth against the action of gravitation, and when the state of strain is
released, energy is developed as the stone
falls.

Another illustration. Suppose a loco-
motive in motion, with full steam up.

Now, let the steam be shut off and the
brakes applied. The engine does not stop
at once, because it has kinetic energy, that
is, energy due to its motion. It does not
stop until the whole of this energy has
been expended in heating the brakes and
the rails.

When it is at a standstill, is all the energy
expended? No, it is not, for even without
burning any more coal, we have only to
turn on the tap, and the *potential* energy
of the compressed or strained steam in the

boiler is released, and, as the engine starts, is changed from potential into kinetic energy, and the motion continues until this again is expended in heating the brakes and rails.

Well, as I said, in the luminiferous ether, when carrying light vibrations, these two sorts of energy exist. The ether is in rapid vibrating motion, so has kinetic energy. It is also in a state of strain, so has potential energy. Note that I only tell you this, and have not given you any proof. The proof is a complex bit of mathematics.

The electric ether is also a receptacle of the two forms of energy, potential and kinetic; and here we can actually partly separate them and study them apart, for when static induction or the induction of electricity at rest is going on, the ether is strained, but is not being kept in motion; while when electro-magnetic induction, that is, the induction of flowing currents of

electricity is going on, motions are produced and kept up in the ether, that is, it
receives kinetic energy.

In the second case, however, it receives
potential energy also.

Possibly, the fact that in the electrical case we can partly separate the two
forms of energy may, at some future time,
throw light on the distribution of the
potential and kinetic energies in the optical
case.

There is another very important point of
resemblance between electric and electro-
magnetic induction, on the one hand, and
light on the other.

It is this,—

In light it is known that the waves are
at right angles to the direction of the
ray.

  I mean that if a ray of sun-light falls
vertically on the earth, its vibrations are
all horizontal.

You know that in this it differs from sound, as the vibrations of the air which are sound are in the same line as that in which the sound is travelling.

It becomes of great interest to determine in which direction the electric disturbance takes place. If I hold this rod over this electroscope so that the line of force acts vertically downwards, then are the vibrations of the ether vertical or horizontal?

Professor Clerk Maxwell has mathematically investigated this point, and has shown that the disturbances both of electro-static and electro-magnetic induction exactly agree with those of light in this respect, for they are both at right angles to the direction of the ray of electric or magnetic induction.

Further, he has shown that, if electro-static and electro-magnetic induction take place together, the electric disturbance will always be at right angles to the magnetic

one ; that is, if the direction of the induction be vertical, the direction of the disturbances will be horizontal, and if the direction
of one of these horizontal disturbances (say

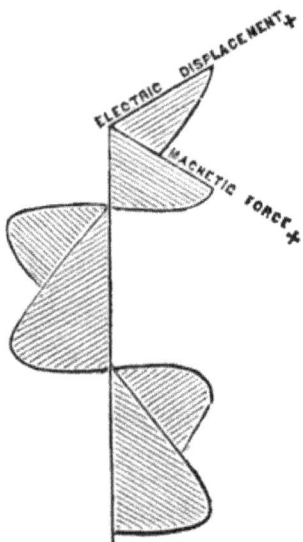

FIG. 33.

the electric one) be from N. to S., that of
the magnetic one will be from E. to W.

This diagram (Fig. 33) is from Professor
Maxwell's book.*

* "Electricity," vol. ii. p. 390, fig. 66.

Another argument in favour of the theory is that it gives a real mathematical reason for the fact that all good true conductors are exceedingly opaque. All metals, for instance, conduct, and are opaque. The conduction of electricity by transparent liquids takes place in a different manner from the conduction by metals, and does not affect the deduction, which can be shown mathematically to be a necessary consequence of the theory, namely, that all good true conductors must be opaque to light.*

Now comes the question, What properties common to both the electric and optic ethers can we observe and measure so as to accurately compare them ?

The first property we shall consider is the velocity with which waves travel in each case.

* It must, however, be confessed that gold, silver and platinum, when made into very thin plates, are not nearly so opaque as they should be according to the theory.

I

I mean we will compare the velocity with which light waves and waves of electro-magnetic induction move in air and in empty space.

In both cases a disturbance is propagated through the ether.

If a candle sends light to the eye, the disturbance—the wave, that is—travels over the space between.

Again, if an electric current by induction affects a magnet at a distance, the strain, or wave, or whatever it is, travels over the space between. With what velocities do the disturbances travel in each case? If these velocities can be measured, and if they can be shown to be the same, it will be a very strong argument for considering that the electric and optic ethers are identical, for the velocity with which a wave is propagated in a medium is a measure of its density and its elasticity.

We shall then consider the velocities

in other media. Both light and electro-magnetic induction are propagated with a different velocity in glass and other transparent solids and liquids to that which they have in air. If these velocities in glass, &c., still agree with each other, we shall have a still stronger reason for supposing that the ethers are not two, but one.

Before we go any further I should just like to remind you about what these velocities are which we are talking of measuring. That of light, you know, is about 185,000 miles per second, or it takes about eight minutes to pass through the 92,000,000 miles between us and the sun.

The velocity of light has been measured directly in many ways. I can only indicate one of them here. We see moving bodies such as planets, &c., at any instant not in the position which they are at that instant, but in the position in which they were when the light left them. Thus, if a planet or star crossed the direct line through a telescope exactly at noon, it would not be seen in the telescope till some time after noon. The length of this time would depend on the distance which the light had to travel.

Now, the earth goes round an orbit 184 million miles across, and, therefore, the light from some stars has to travel 184 million miles further at one part of the year than at six months later. If the instant at which one of these stars is seen to cross a fixed telescope is noted, and the same observation is repeated six months later, there will be found to be a difference of about sixteen minutes, which difference is due to the fact that at one time the distance from the star to the earth is longer by some 184 million miles than at the other. From exact measurements of this time the velocity of light has been determined.

Many other methods of measurement

have been used; in one of which the time required by light to travel fourteen yards was determined, and the result agreed very well with the one where the distance was the diameter of the earth's orbit.

The results obtained by different observers are in the left hand upper column of the table (p. 118).

The velocity of electro-magnetic induction has not yet been measured directly.* Probably no attempt at a direct measurement will ever be made, for we have an indirect method of computing it, which is susceptible of far greater accuracy than is ever likely to be obtained by a direct measurement.

A comparison of electro-static and electro-magnetic actions is the basis of this measurement. I fear the process is too

* The nearest approach to a direct measurement has been made by Mr. Rowland and Prof. Helmholtz, Phil. Mag., Sept., 1876, p. 233, who measured the velocity with which a body charged statically must be moved to produce a certain magnetic effect.

VELOCITIES.

MEDIUM.	VELOCITY OF LIGHT.		VELOCITY OF ELECTRO-MAGNETIC INDUCTION.	
	Authority.	Miles per Second.	Miles per Second.	Authority.
Air and Vacuum .	Fizeau . .	195,100	193,000	Weber
	Astronomi-cal obser-vations . } 193,300		187,900	{ Rowland & } Helmholtz
			178,900	Maxwell
	Foucault .	185,300	175,200	Thomson
	Cornu . .	187,100	183,100	M'Kichan
			185,300	{ Ayrton & Perry
	Mean . .	189,700	183,900	Mean
		186,000		
Double extra dense flint glass . . .	Gordon {	106,400	104,600	Gordon
Extra dense flint glass . . .		110,900	106,400	
Light flint glass .		123,100	107,300	
Hard crown glass		116,800	105,500	
Paraffin . . .	Gladstone and Maxwell . }	130,800	131,700	
Plate glass . . .	Text Books	121,000	103,000	
Sulphur		89,000	115,000	
Bisulphide of Carbon . . }		115,000	139,000	

purely mathematical to enable me to explain it in detail, but I can indicate the principle of it. If the same thing, such as the resistance which a wire offers to the passage of electricity, be measured both electro-statically and electro-magnetically, the numbers obtained for the result will not be the same.

The difference is caused by one set of measurements being based on the consideration of electricity at rest, and the other on the consideration of electricity in motion. Mathematics tell us that the ratio of the results is a motion or a velocity, and that this velocity is the velocity of electro-magnetic induction; therefore, from such measurements this velocity can be calculated. A comparison of the two upper columns of this table shows a very close agreement between the velocities of electro-magnetic induction and of light. You notice that the difference of the means of

the two columns is less than the accidental
differences between numbers in the same
column.   Thus we have seen that *in air
and vacuum the velocities of light and electro-
magnetic induction are sensibly equal.*

Now we will consider the case of glass
and other transparent insulators.

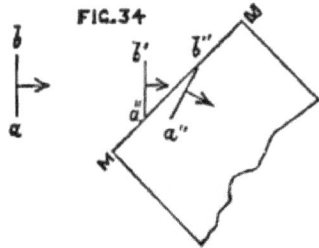

FIG. 34

In these light does not travel with the
same velocity as in air.

How do we determine the velocity of
light in glass?   We do not determine it
directly, but measure the difference of velocity in air and glass.   Let *a b* (Fig. 34) be
the front of a wave of light travelling along
in the direction of the arrow.   Let it fall on
a piece of glass M M, placed diagonally

to the direction of the ray, or arrow.
When the wave gets to the position a' b'
the lower part a' begins to enter the glass,
the upper part b' is still in the air. But
as light travels more slowly in glass than
in air, by the time the top b' has got to
the glass at b'' moving in air, the lower
part a' moving in glass has not travelled
so far as the upper part, and has only got
to a''. The lower part of the wave is
retarded or dragged back behind the upper
part and the wave, and, consequently, the
direction of the ray is twisted round, so as
to make an angle with its former direction.

Suppose two people are pushing a two-
wheeled cart along by turning the wheels.
If one turns his wheel faster than the
other does, the direction in which the cart
travels will be changed.

Now, if we measure the angle through
which the cart turns, we can calculate the
difference in the speeds at which the two

wheels are moving. Similarly, if we measure the angle through which our ray of light is turned, we can calculate the difference of the velocities with which the part in air and the part in glass were moving.

The lower left-hand column of the table (page 118) shows the results of the calculations for the velocities of light.

The velocities of the electro-magnetic induction are calculated from the velocity in air.

Theory shows that this calculation can be made when we know the specific inductive capacities.

The greater the specific inductive capacities the slower the electro-magnetic induction travels.

The lower right-hand column of the table (page 118) shows the results of the calculations for the velocities of electro-magnetic induction.*

* The velocities in the lower half of the table are

You see that in certain dielectrics there is a very close agreement indeed, notably in paraffin and in some of the denser

calculated from the refractive indices and specific inductive capacities given in the following table :—

Dielectric.	$\sqrt{K}$	Nearest value of μ.	Ray for which μ is nearest.
Double extra dense flint glass	1·7783	1·7460	Band in extreme violet in magnesium spark spectrum.
Extra dense flint .	1·7474	1·6757	
Light flint............	1·7343	1·5113	
Hard crown	1·7629	1·5920	
Plate glass	1·8009	1·543	
Paraffin	1·4119	1·422	Rays of infinite wave length.
Sulphur..............	1·6060	2·115	
Bisulphide of carbon................	1·3456	1.6114	

Here K is the specific inductive capacity, and $\sqrt{K}$ stands for the square root of K. μ is the refractive index.

Maxwell's theory gives that the velocity of electric induction in any dielectric is to the velocity in air glasses; in others there is a very wide difference.

I am hoping shortly to make some further experiments, both on sulphur and on bisulphide of carbon. I am not certain of the accuracy of my determination of the specific inductive capacity of bisulphide, as liquids present special difficulties; while as to sulphur, it has so many different forms —yellow crystal, red powder, black plastic

inversely as the square root of the specific inductive capacity. To determine the velocity of electric induction in any dielectric 186,000, the adopted mean velocity in air, is divided by the number under the heading $\sqrt{K}$ in this table.

We also know that the velocity of light in any medium is inversely as the refractive index μ. Hence the velocity of light in each dielectric is found by dividing 186,000 by the number under the heading μ.

The values of $\sqrt{K}$ are from the author's determinations of specific inductive capacity (page 90). μ was determined by the author for the first four glasses; for paraffin the value given is that calculated by Maxwell from Gladstone's experiments. For the other substances, where there is a wide difference, the values of μ are taken from the text-books.

wax—that it is possible that the discrepancy may be partly accounted for by supposing that there was a difference in the physical state of the sulphur when the electrical and optical experiments were made.

At present, however, I think we may fairly say that in some dielectrics the velocity of electro-magnetic induction is nearly equal to the velocity of light. That there is almost always a small difference, and that sometimes there is a very large difference.

That it is quite possible that the relation which we have spoken of between electric induction and light exists, namely, that they are disturbances of the same ether ; but that there is some unknown disturbing cause affecting the electric induction, and that in some dielectrics the disturbing cause is very small, but that it is in others large enough to cause a very large difference between the velocities of light and the calculated velocities of electro-magnetic induction.

I am hoping some day to compare the electro-static inductions along and across the axis of the same crystal.

I hope that the disturbing cause, whatever it may be, may affect the inductions equally in both cases.

We know that the velocity of light in a crystal is different along and across the axis, and if then the ratio of the two light velocities were the same as the ratio of the two electro-magnetic velocities, we should have a confirmation of the theory, independent of any knowledge of the nature of the disturbing cause.

What this disturbing cause may be we do not know. Perhaps some future investigation may explain its real nature.

I am now going to attempt to show you a few experiments to illustrate other connections between electricity and light, experiments in which electricity acts on light and *vice versâ*. In fact, without too bold a hypothesis, we may call them experiments where, on the ether in certain bodies being disturbed by electricity, the disturbance is seen in their changed action on light, and where, when it is disturbed by light, their action on electricity is altered.

The first action which I will show you is the electro-magnetic action discovered by Faraday.

Faraday found that if an electric current were made to circulate round and round a cylindrical ray of light, that, *in certain media*, the ray would be twisted, so that a line drawn along the outside of the ray would no longer be straight, but would be twisted spirally like the rifling of a gun.

But how are we to draw this line, and how see the twist of it ?

You know that in ordinary light the

vibrations take place in every possible
direction at right angles to the ray.
Crystals of Iceland spar, cut and arranged
in a particular way, called " Nicols' prisms,"
have, however, the power of compelling all
the vibrations to take place in one particular
plane, or of polarizing the light, as it is
called. In fact, when the light is plane-
polarized, we have a flat ray of light
instead of a cylindrical one. Let us now
pass light horizontally through a Nicols'
prism, so as to polarize it, let us say, in a
horizontal plane.

We shall now have extinguished all
vibrations, except those in one plane, say
the horizontal plane.

Let us now put a second prism, with its
polarizing plane vertical; it will have the
power of extinguishing all horizontal
vibrations, and it will, therefore, entirely
extinguish all the light which has come
from the first prism, as you see. But if

the plane of the light is twisted by any means
between the two prisms, it will no longer fall
with its vibrations horizontal on the second
prism, but will be partly allowed to pass, more
and more of it being admitted as the plane
is twisted more and more nearly vertical.

Here, between the prisms (Fig. 35), is a

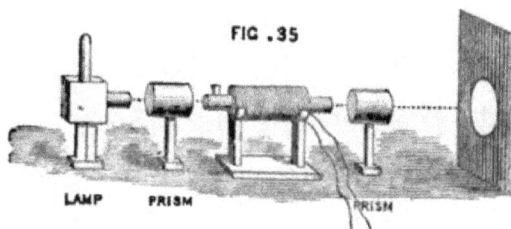

FIG .35

LAMP   PRISM   PRISM

tube with glass ends, filled with bisulphide
of carbon, and here is a coil of wire of 1028
turns, carrying an electric current round
and round the tube. The light passes along
the tube. The prisms are set to extinction.
I turn on the electric current of ten Groves
cells, and you see the re-appearance of the
light on the screen shows that the plane of
polarization has been twisted. I have to
turn the prism through about 14° to again

K

extinguish it. The number of degrees which
the prism has to be turned of course show
the amount of twist that has been given to
the light. On reversing the current the light
re-appears, and to extinguish it we have to
turn the prism as far to the right of the
zero as we previously turned it to the left.*

Now, as many of you will know, a
spiral current acts just like a magnetized
bar would if placed in the spiral. In fact,
if light is sent from pole to pole of a
magnet, the same effect is produced on it
as that which we have just seen. In 1877 a
paper of mine was published in the " Phi-
losophical Transactions,"† containing measurements of the amount of twist which a
unit magnetic force would give to a ray of
light, and in the " Comptes Rendus " for

* *See* Faraday, " Experimental Researches," § 2146,
vol. iii. ; and Verdet, Œuvres, T. i., Notes et Mémoires.

† Vol. 167, p. 1, " On the Determination of Verdet's
Constant or Absolute Units."

1878,* M. H. Becquerel pointed out that from these results we can calculate what would be the effect of the earth's magnetism on light in certain media.

If a canal one mile long were dug from north to south near Kew, and filled with bisulphide of carbon, a ray of green polarized light entering at one end, would, by the action of the earth's magnetism, have its plane of polarization twisted just 50°. There are slight differences in the action on different coloured lights. My measurements were made on the green light of burning thallium. If the canal had been full of distilled water the twist would have been about $7\frac{1}{2}°$.

The explanation of this phenomenon is still exceedingly obscure. We as yet know so little about the molecular structure of bodies that there are very many gaps in the chain of reasoning, that must either

* T. lxxxvi. p. 1077.

still be left empty or filled in with provisional hypotheses, that is guesses.

What we do know on this subject may be briefly summed up as follows: In the disturbance which we call light, whatever its true nature may be, we know that there is something like a rotation round an axis going on, which axis is the direction of the ray.

When magnetic forces act on a medium, Professor Maxwell has shown that there is always something like a rotation round an axis going on, which axis is the line of force. But here the resemblance stops. There is nothing in the magnetic phenomenon which corresponds to the wave length and wave propagation in the optical phenomena.

As to the nature of the rotation accompanying the magnetic forces, we know that it exists, and we know that it is not the rotation of any sensible portions of the medium as a whole.

Professor Clerk Maxwell suggests that it may be a rotation of "molecular vortices," that is, that every part of the magnetized medium may be filled with little whirlpools exceedingly minute. These whirlpools may be motions of ultimate particles of matter, or may be motions of the ether in it, or possibly in this region of "very-smallness" the ether and the matter may be one. Now, as minute eddies in a stream whirl chips round but do not affect a large boat, so these whirlpools, while they cannot affect the sensible motions of bodies, may be able to influence greatly the minute vibrations which are the propagation of light.

The next experiment I have to tell you about is on the action of electro-static induction upon light. You know that if light is sent through a crystal it is acted on in a way different to the action which occurs when it passes through a homogeneous medium. In particular, if light is plane polarized before it enters the crystal —that is, if its vibrations are all in one plane—then, after it emerges from the crystal, some of the vibrations will be circular, and in no position of the second Nicols can the light be extinguished.

In Nov. 1875, Dr. Kerr* announced

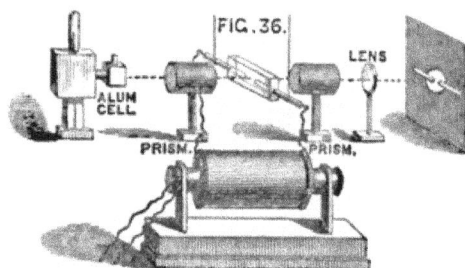

FIG. 36.

that when glass is subjected to an intense electro-static strain it acquires the same action on light as a crystal has. In fact, that the electric strain so rearranges the molecules of the glass that they act on light as if they were the naturally-arranged molecules of a crystal. Here is the method

* "Phil. Mag.," 1875, part ii. p. 337.

he used (Fig. 36), but the effect, though quite decided, is so minute, and the conditions of success are so delicate, that I have not much hope of showing the actual experiment to you.

Here is a piece of thick plate.glass, about eight inches long and two wide having two holes drilled in it from the ends, so that they come within about 3-16th of an inch of each other. (Fig. 37). Into these holes wires

FIG . 37

are cemented. These wires are connected to the secondary poles of the large induction coil, which we used last lecture (see Fig. 23), only now it is used with its full power, with ten quart-sized Grove cells, and its own vibrating break. When the coil is worked, the tension across the 3-16th of an inch of glass is, of course, equal to the tension across the air between the discharging poles where the sparks are passing. By commencing with the poles close together, and then gradually drawing them apart, the tension across the glass can be increased as we please. The Nicols prisms and a lens for focussing are arranged, as shown in Fig. 36. An alum cell is attached to the electric lamp to intercept the heat rays. Now, we turn the Nicols, so as to darken the screen, start the coil, and gradually draw the discharging poles apart. When we get to a tension equal to about seven inches of air you see the patch of white light* appearing on the screen. (This was clearly seen by a large audience.)

* In rehearsing this experiment the day before, Mr. Cottrell, Mr. Valter (the second assistant), and myself only being present, the strain was accidentally allowed to become too great, and the glass was perforated. Immediately before perforation some extraordinary effects were seen on the screen. First appeared a patch of orange-brown light about six or seven inches diameter. This at once resolved itself into a series of four

Dr. Kerr found that the maximum effect was produced when the prisms were so set that the line of electric strain was at 45° to the direction of optical vibration. (Fig. 38.) As the prisms are turned away from that position the effect gets less and less, till, when the direction of vibration is either

or five irregular, concentric rings, dark and orange-brown, the outer one being, perhaps, fourteen inches diameter. In about two seconds more these vanished and were succeeded by a huge black cross about three feet across, seen on a faintly luminous ground. The arms of the cross were along the planes of polarization, and, therefore (the experiment being arranged according to Dr. Kerr's directions), were at 45° to the line of stress. The glass then gave way, and all the phenomena disappeared except the extreme ends of the cross, and the discharge through the hole, where the glass had been perforated, was alone seen. I have since made numerous attempts to repeat this effect, but have not succeeded in doing so, though I have perforated many valuable glasses. In this particular case the glass happened to break slowly. In all the repetitions of the experiment the glass has broken suddenly, and there has been no time for the new effects to occur. Proc. Roy. Soc., Feb. 13, 1879.

along or perpendicular to the line of strain, there is no effect at all.

a a line of electric strain.
b b, b b' direction of optical vibrations.
Ray of light perpendicular to plane
of paper.

We observe that in this experiment there is no rotation of plane-polarized light as in the last one, for no rotation of the Nicols will extinguish the light. It is found that the light after emerging from the strained glass, is no longer plane-polarized, but that its vibrations are circular.

I have still one more experiment to show you.

Here is a metal called selenium. It conducts electricity, but very badly—that is, it offers a great but not an infinite resistance to the straining force.

But it has this extraordinary property.

It conducts much better in the light than
in the dark.   The light vibrations actually
seem to shake its molecules, and help them
to yield to the electric strain.

I will try to show you this experimentally.

Here you see the arrangement. (Fig. 39.)

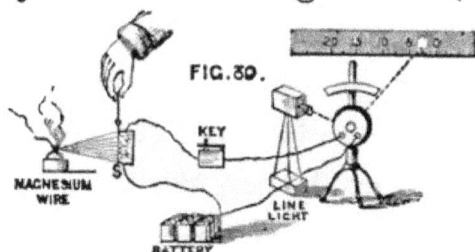

FIG. 39.

Here is a delicate reflecting *galvanometer*
of high resistance.   On pressing down the
contact key the current from a ten-cell
Leclanché battery flows through the galvanometer, and through a piece of selenium
enclosed in a light-tight box, and you see
the deflection of the needle moves the spot
of light over about ten divisions of the scale.
Now, we light this piece of magnesium
wire, and draw up the sliding front of the

box. The deflection is at once doubled, showing that in the light this selenium conducts about twice as well as in the dark.

Prof. W. G. Adams has made many experiments on this subject,* and he found that light can actually *produce* a current of electricity, and not merely aid its passage.

Here † is a box containing a piece of selenium whose ends are connected to the galvanometer. I open the box and admit the light of the electric lamp and there is a deflection of the galvanometer.

The electricity of a battery has been converted into light in the lamp, and that very same light is again converted into electricity in the selenium.

In these four lectures we have considered some few phenomena which we can explain,

* Proc. Roy. Soc., 1876, xxiv. p. 163.
† This last experiment was omitted for want of time.

and a great many which we cannot. Most of the experimental facts stand as yet alone and disjointed. Many lines of reasoning and research open out a little way, and then are lost in the darkness, or, rather, let us say in the brightness through which, as yet, human sight cannot pierce.

No doubt the day will come when all these difficult ways will be clear and trodden paths, when all these disjointed facts will be seen to be parts of one true, harmonious, and perfect whole.

THE END.

INDEX.

www.ingramcontent.com/pod-product-compliance
Lightning Source LLC
Chambersburg PA
CBHW021813190326
41518CB00007B/577